Published by
Brown Paper Package Publications
Carlinville, IL 62626

ISBN: 0-9773635-0-3

Printed in U.S.A.
September 2005
First Edition

Cover and Book Design by Laurie A. Flori

"A Dedication"

Oh Roseberry, Roseberry
With cracked plaster walls
And rusty old gutters
Twas you we picked
From a hundred fifty-one others.

We knew you were ours
With your steep gable dormers
And strangely slanted ceilings
In all the upstairs corners.

But, alas Roseberry, Roseberry
You really needed work
And lots of T. L. C.
We fell in love with you
And the rest is history.

You were a swan
Amidst all the ducks
We knew to restore you
Would cost a few bucks.

Yes Roseberry, Roseberry
We fixed every cranny
And your every single nook
Took you back to the grandeur
Of how you looked in the book.

In the time we've restored you
The twenty-four plus years
You've shared our laughter
And cried with our tears.

Thanks Roseberry, Roseberry
For welcoming our babies
You didn't bat a door
You kept them safe
Gave them a home forever more.

We know your inner secrets
Every cracked pre-cut board
There's a bond between us
That cannot be ignored.

Oh Roseberry, Roseberry
Our Home Honor Bilt
No matter where we're bound

Tis you we ache for
When the chips are down.

You've stood tests of time
Your hundredth birthday's near
Hope we look as good as you
When near our hundredth year.

So Roseberry, Roseberry
Tis to you I dedicate this book
Every line on every page
I promise as long as I live within you
To never, ever let you show your age.

You're our happy much-loved Mansion
And when we reach heaven
God will surely have awaiting us, a home,
Stamped Sear's style 2037.

Happy Hundredth Birthday Roseberry!

Our Roseberry

"How I Became Obsessed"

How I Became Obsessed

*Just how did I become obsessed with the Sears Homes and Standard Addition in Carlinville, Illinois? That is a good question. I have to think back to the days when my husband and I were first married; we were married on **October 2, 1981.** A couple of years later, we wanted to move from our small apartment to a house. It just so happened that a friend of my husband's family, Mrs. Gloria Beasley Carriker, owned several homes in Standard Addition that were for rent. We went to see the house and immediately loved it. We moved in. My husband had grown up in the Addition on Charles Street where his parents had owned their 1st home, and he said it felt like he was coming home.*

We found that we were expecting our 1st child, Jacob, a couple of months after moving into the house. After Jacob was born, we asked the owner of the house if she was interested in selling and she was. We purchased it for $18,000.00 – it was a steal. I had all of these dreams of what I wanted to do to restore the house and immediately started work in the living room and dining room. By the time we had restored those rooms to what I thought they should look like and were well underway in the kitchen, we found out we were expecting our 2nd baby, Christopher. So we just superficially finished the kitchen, with the original kitchen cabinets and the bathroom with the original sink and skirted tub and focused on the boys.

*Time went on and the kids were our lives; we spent all the time we could with them. You could count the babysitters we had on one hand. I had just finished hanging major league baseball wallpaper up in the boys' bedroom when a friend, Danley Vlasich, asked if we would consider opening our house for a tour **September of 1986,** to launch the "1001 Home Ideas" issue that featured a Sears, Roebuck & Co. Honor Bilt Home in Standard Addition. This feature was to come out in the next summer's issue of the magazine. The home had been totally gutted and revamped by Sears, complete with antique furnishings. The magazine was going to have an open house to promote it. Danley, as well as officials from Sears, went through our home and asked again if we would consider opening it. My husband and I thought it over and thinking that maybe there would only be a couple of hundred people through our home on that Sunday; we said yes.*

Well, with two boys, one not even a year old, and the other not even two years old, we set about the task of cleaning and getting ready for the tour. Needless to say, the boys spent a great deal of time at Grandma Flori's while we readied the house.

The tour day came; it was a beautiful Sunday afternoon and very warm. We had friends in every room of the house and my friend Teri Cerar was the ticket puncher on the front porch. You were required to buy a $3.00 ticket and that got you in to see all of the homes that were open, including the 1001 Ideas redone one, as well as many of the gardens in the Addition. We couldn't have been prouder and the headline in the paper read, "Standard Addition Sparkled" and it certainly did that day.

Instead of the 200 people that we thought would venture through on that day, there were more than 2000 – that's right, two thousand in three hours on a Sunday afternoon. We were shocked; the line of people to see our house stretched clear to the end of the block more than 7 houses away, and there are only 9 homes on every block. Teri said that her hand hurt from punching so many holes in so many tickets., Those who saw every house and therefore had every hole punched could enter a big drawing for prizes at the end of the day. By the end of the afternoon, my voice, as well as my husbands, was totally gone from explaining the Standard Addition and Sears, Roebuck & Co. Homes Saga to all of the tourists. All of those tourists listened earnestly to every detail about Standard Addition and our house. They were totally fascinated with the story. It was then that I knew, it wasn't just my house I loved, it was the whole story surrounding it; the fact that 2000 people on a Sunday afternoon came to see it, the fact that Sears still cared about them and most of all, people from everywhere were interested in seeing them. I had to do something to preserve the Sears homes in my community.

I started having tours. Busloads would come. It didn't matter if they had money to buy tickets or not; we wouldn't charge them. People would drive up in the afternoon or on a weekend and say, "I read about your house and just wanted to see it. Do you mind if we take some pictures?" We'd say, "Sure, go ahead. You can go in if you don't mind the fact that we are in a remodeling project. The house hasn't been cleaned." You know what? People didn't care; they still don't. They just want to hear the history and to say they've been here.

I became totally obsessed. Since then we have hosted many, many tours, magazine shoots, T.V. crews, news shows, T.V. shows, documentaries, authors, people writing essays, themes, dissertations and more. We still have the people who pull up on a weekend and say, "I was in town and just thought I'd come by and see your house," and you know what, we still let them in – we seldom say no.

We have finished almost all of our restoration projects; there are still some things that we would like to see done, like my husbands hockey arena, a skating rink, my turret on the house, my wish for a wide open floor plan, a built in pool (the above ground one we have is beautiful and it will have to do), a weight room, a jacuzzi room, but that is literally – a whole new house.

Through the years we have seen families come and go. We've had more people who have moved into the Addition to make it their home as we have, but we still have the ones who don't care - they move into a house so they can get cheap rent. Some people buy some of the houses and do nothing to them. They rent them out and become slum landlords. These people break my heart. The local real estate agents are the ones who really make me angry, they call them "just Standard Addition" houses or "starter" houses – what is meant by a starter house anyway? I thought they were supposed to promote the good things about a house.

I will never stop fighting for things in Standard Addition. I will never stop trying to keep it cleaned up. Most of all, I will never stop trying to get it recognized as a place of historical and honorable significance. If my house were to ever be listed by a real estate agent, they better not ever call it "just a Standard Addition" house or a "starter" house, because what my family and I live in is a **Genuine Sears, Roebuck & Co. Honor Bilt Home called the Roseberry Model in the middle of the largest single collective of Sears, Roebuck & Co Homes on earth, and I am proud.**

And yes, darn it - **I AM OBSESSED!**

A
"Special Thank You"
to my husband David,
our children Jacob and Christopher
and the family pets - Barfbag, Pounder and Buddy,
for all of the years they have tolerated, helped, consoled and been there for me through my obsession.

"Additionally Speaking"

By Laurie A. Flori

A chronological history of the Sears, Roebuck & Company Homes Saga in Carlinville, Illinois.

Index

Special Thanks to my former bosses, Ed Albracht and Thomas Hatalla, editors and former owners of the **Carlinville Democrat** Newspaper, for use of the **Carlinville Democrat** Archives and for giving me permission to use all of the information that I wanted from the old newspapers. I could not have written this book without their generosity.

Chapter I

"From the Start"

From The Start

☙ The Beginning ❧

❧ The Saga of Carlinville, Illinois Historic Standard Addition begins way, way, way back in the fall of **1906**…

…Three strangers were seen walking through the Mississippi bottomlands. Years earlier, Lewis and Clark had set up camp in the very same area just before leaving on their legendary journey. This time, the legendary journeymen were a Dr. Burton, W.P. Cowan and Mr. J.P. Evans of the Standard Oil Company of Indiana.

These three Standard Oil Officials had gone to the area to scout out a possible location of a new refinery site. This was a prime location, known as Wood River Valley, and was situated between the bluffs and the Mississippi River, through which the Wood River flowed. Except for a few farming areas and the railroad line that ran through the center, it was virtually uninhabited.

Note: ***If everyone knew then what we know now, these events would most surely have been documented in the greatest of detail, but we don't always know that something is important – really important – until after the fact.***

Wood River Refinery in ***1907.***

The officials immediately began purchasing up the land on both sides of the railroad line at $200 to $500 an acre. It was not very valuable land to anyone else and therefore a shrewd investment of $93,000 for Standard Oil. The land west of the railroad line gave the company a frontage of 5,700-feet on the Mississippi while a larger tract, east of the railroad, gave them the area needed for the refinery site. Construction on the new refinery began in the spring of **1907**, and by late that year, the refinery was completed. E.M. Clark was manager of the refinery.

By **1912**, Standard Oil had increased its facilities to 1,407 sites across Illinois, Iowa, Kentucky, Indiana, Michigan, Missouri and Kansas. In the years that followed, Standard Oil opened facilities all over the Middle West.

*Right is a **1906** Chicago & Alton Railroad Train.*

*Below is the Standard Oil Company Refinery at Wood River, Illinois in **1908**.*

We Need Coal

In **1917**, one of the company's biggest difficulties was finding an adequate supply of coal to power its burners beneath the large stills in which gasoline was refined. It became so difficult to get coal that when a supply was found the cost was outrageous.

Special Thanks to Thomas R. Giddens, Jackson A. Giddens & Judy (Giddens) White for use of the images on this page, from their father, Paul H. Giddens' book "Standard Oil Company of Indiana - Oil Pioneer of the Midwest."

Standard Oil Company Office in Whiting, Illinois.

E.M. Clark – still the manager at the Wood River Refinery – proposed that the company buy a coal mine and mine its own coal to offset these costs and this shortage. The directors liked Clark's idea and started an investigation into the possibility of owning their own coal mine.

After the exploration into several minefields, an area near Carlinville, IL was chosen. It was a great location, as it was located between Whiting, IL (another refinery site) and Wood River, IL. The freight rates would also be cheaper in this area because it was right on the main line of the Chicago & Alton Railroad from St. Louis to Chicago. They began drilling several test holes in the northeastern part of Macoupin County. These test holes indicated a very good seam and grade of coal. Standard Oil officials immediately purchased the underlying coal rights to this and the surrounding lands.

Right is the Standard Oil Company Refinery at Whiting, Illinois.

*Below is the No. 360 Chicago & Alton Train that ran through Carlinville, from **1909 to 1935**.*

The Carlinville Coal Mine Is Sold

THE COAL MINE SOLD

OWNERSHIP OF BIG INDUSTRY OF CARLINVILLE CHANGES.

Sale Made Last Week—Purchasers Take Over Property—Deal is Probably Beginning of Big Things.

On **October 17, 1917**, the Carlinville Coal Mine was sold.

Note: The Carlinville Coal Mine was sunk in 1867 by the Weer Brothers. The depth from the shaft to the top of the vein was 266-feet. The vein itself was 6-feet in depth. The mine hoisted about 500,000 bushels of coal a year and employed around 45 men.

Carlinville property. The deed of transfer is made to Thomas Williamson, as trustee. Mr. Williamson is an Edwardsville lawyer. Neither he nor any of the parties to the deal have announced just who the purchaser is. The fact, however, that Mr. Williamson has been and is representing the Standard Oil company in large matters, leads many to believe that the great oil company is the real purchaser of the local coal mine.

At the time of the purchase it was speculated that the purchaser was Standard Oil of Indiana because Thomas Williamson was the trustee and he was an Edwardsville lawyer, who had represented Standard Oil in such affairs, and he, himself, completed the negotiations for the purchase and transfer of the Carlinville Coal Mine. The completed sale included the mine, buildings, equipment and coal lands owned by the Carlinville Coal Mining Company. The purchase price was $25,000. One of the men who also took part in this transaction was E.M. Clark. It was reported that he had been in Carlinville more than once to take part in negotiations.

On **October 24, 1917**, in the **Carlinville Democrat**, it was reported that the purchaser of the Carlinville Coal Company was indeed Standard Oil of Indiana.

Left, is the old Carlinville Coal Mine, which was located near the railroad crossing on West Nicholas Street. The building on the left still stands there today and is used by the Carlinville Street Department.

Operators Robert Leon and H.T. Simpson, shown with the coal cutting machine in the Carlinville mine in ***1922.***

Carlinville miners waiting to get on the cage to go into the mine in ***1922.***

Right, are Philip Ward and Mike Corkin, operaters of the rotary dump in the Standard Oil Mine in ***1922.***

Colonel Stewart, members of the board and others inspecting the Carlinville Coal Mine. Standing L to R – C.W. Clark, Edward G. Seubert, W.M. Burton, Charles Manbeck, Colonel Robert W. Stewart, T.J. Thompson, W.E. Warwick and Tom Paul. Kneeling L to R – John McMillan, Allan Jackson, E.J. Bullock, R.H. McElroy and Rollin Smith.

Carlinville Coal miners practice with a breathing apparatus for use in mine rescue in early ***1900's.*** *(L – R) Steve Fontilla, Andy Peters, T.W. Starbuck, Ralph Milhors and John Gemp.*

No.701 Chicago & Alton Railroad Train that ran the Alton/Chicago corridor in ***1914.***

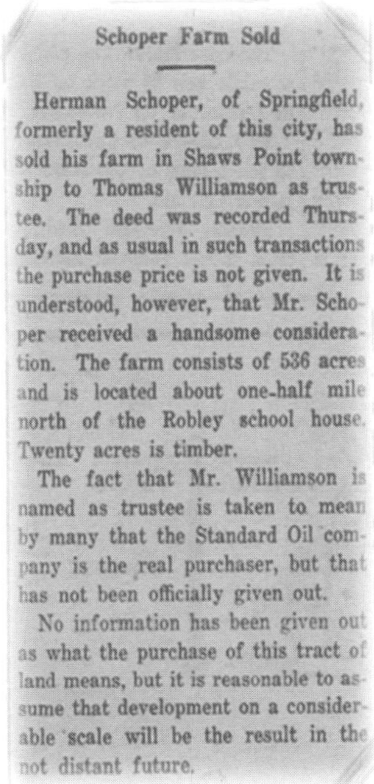

Schoper Farm Sold

Herman Schoper, of Springfield, formerly a resident of this city, has sold his farm in Shaws Point township to Thomas Williamson as trustee. The deed was recorded Thursday, and as usual in such transactions the purchase price is not given. It is understood, however, that Mr. Schoper received a handsome consideration. The farm consists of 536 acres and is located about one-half mile north of the Robley school house. Twenty acres is timber.

The fact that Mr. Williamson is named as trustee is taken to mean by many that the Standard Oil company is the real purchaser, but that has not been officially given out.

No information has been given out as what the purchase of this tract of land means, but it is reasonable to assume that development on a considerable scale will be the result in the not distant future.

☙ Schoper Farm Is Sold ❧

❧ On **October 31, 1917**, Herman Schoper, of Springfield, IL, formerly of this area, sold his farm in Shaws Point Township to Thomas Williamson as trustee. The farm consisted of a 536-acre tract, which was located about a half of a mile north of the old Robley Schoolhouse in rural Carlinville.

Twenty-acres of the Schoper farmland was timber.

Many people thought since Mr. Williamson was named as trustee of this proceeding it meant that Standard Oil was the real purchaser of this as well. At the time, no one had any idea just what this land was to be used for. Everything was very secretive, and neither Mr. Williamson, Mr. Clark nor anyone else connected with Standard Oil were talking to anyone – yet.

It was assumed by everyone that these purchases meant something of considerable scale was in store for Carlinville.

Whoever was buying up the land – if indeed it was Standard Oil – was buying it up at too fast of a rate for something not to be happening.

But what?

Note: They assumed right. What was in store and eventually took place, was probably the biggest thing that has ever happened to Carlinville and the surrounding areas.

Nothing had ever, or would ever, make such an impact on Carlinville as this had. It's just too bad the events that took place in those years were not documented in great depth and detail because their significance was a change in the Carlinville, Illinois history timeline that was never to be reversed.

It would take almost 85-years for this to become something of historical significance.

Chapter
II

"The Sinking of Two Mines"

The Sinking Of Two Mines

Standard Construction Progressing

The **Carlinville Democrat's, December 13, 1917**, edition reported, *"Standard Oil construction work at the new mine sites eight miles northeast of Carlinville was progressing favorably."* Mullville Bros. were supervisors of the entire construction and had employed workmen to lay tracks from the Chicago & Alton Railroad track connections to the yard tracks at the proposed mines. The paper also said *"Houses were to be built immediately for quarters to board laborers."*

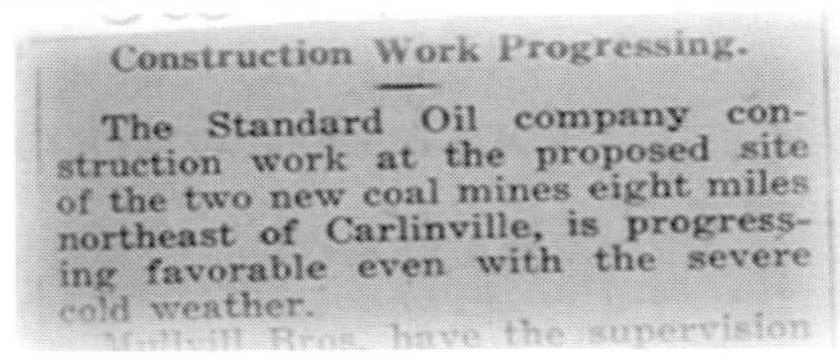
Construction Work Progressing.

The Standard Oil company construction work at the proposed site of the two new coal mines eight miles northeast of Carlinville, is progressing favorable even with the severe cold weather.

Mullvill Bros. have the supervision

A special Thank You to Mary Schroder for the photo of the Schoper Mine under construction.

The Schoper Mine being built in 1919.

On **January 9, 1918**, the **Carlinville Democrat** stated, *"The outlook for prosperity for Carlinville and the community is very bright. The recent purchases by Standard Oil and their recent developments of these properties were a guarantee of the gratifying enlargement of business activities in Carlinville and the area."*

Standard Oil Night

One of the most interesting meetings to ever be hosted by the Carlinville Commercial Club was on **January 18, 1918**. It was said to have been called, *"Standard Oil Night."* Standard Oil Company Officials, who were guests at the Commercial Club Banquet, had discussed *"Carlinville's Future"* that Thursday evening. They outlined their plans and the many things that were to take place in Carlinville. Mr. Woodward, president of the Commercial Club, took the floor and briefly stated that the Commercial Club had invited these representatives to be at the meeting. Thomas Williamson spoke on behalf of Standard Oil regarding their activities in Carlinville. He explained that he represented Mr. Clark who expected to be present, but had been called out of the city on important business. Mr. Williamson talked about what Standard Oil was going to do in Carlinville in the near future by comparison to what they had done in Wood

River the past few years. His intimation was that activities of at least as large a scale would be the result of Standard's coming to Carlinville. He then pointed out all of the great things that had taken place in Wood River since they had opened their refinery there, including: paved streets, electric lights, sewers, two school buildings and more. Wood River now had all of the conveniences of an up to date and modern city; this all was owing to the fact that Standard now paid 90% of all of the taxes that were collected in Wood River. Mr. Williamson then stated, *"The Company was pleased with Carlinville."* His talk had been listened to with the closest of attention, and all that he had said greatly impressed his listeners.

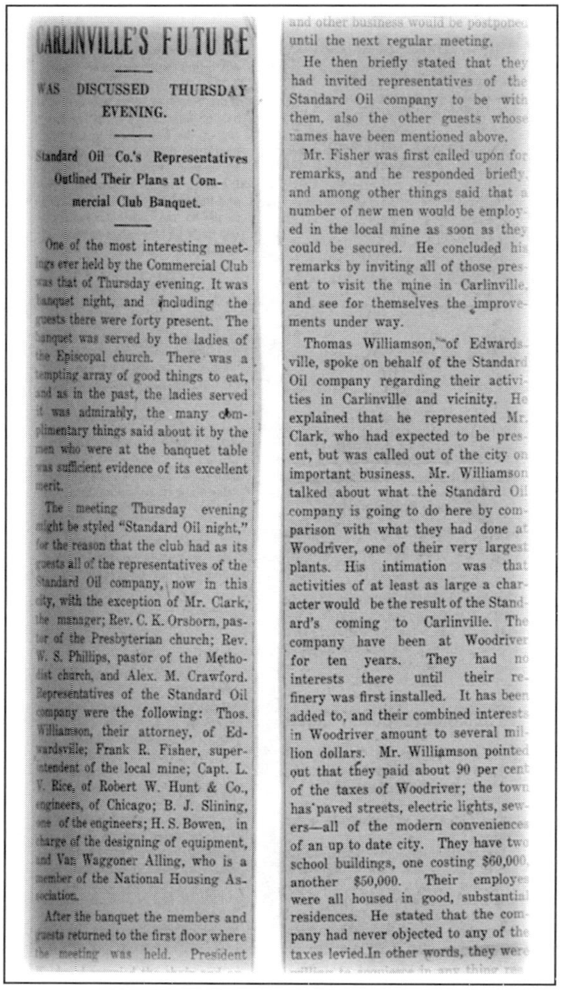

CARLINVILLE'S FUTURE

WAS DISCUSSED THURSDAY EVENING.

Standard Oil Co.'s Representatives Outlined Their Plans at Commercial Club Banquet.

One of the most interesting meetings ever held by the Commercial Club was that of Thursday evening. It was banquet night, and including the guests there were forty present. The banquet was served by the ladies of the Episcopal church. There was a tempting array of good things to eat, and as in the past, the ladies served it was admirably, the many complimentary things said about it by the men who were at the banquet table was sufficient evidence of its excellent merit.

The meeting Thursday evening might be styled "Standard Oil night," for the reason that the club had as its guests all of the representatives of the Standard Oil company, now in this city, with the exception of Mr. Clark, the manager; Rev. C. K. Orsborn, pastor of the Presbyterian church; Rev. W. S. Phillips, pastor of the Methodist church, and Alex. M. Crawford. Representatives of the Standard Oil company were the following: Thos. Williamson, their attorney, of Edwardsville; Frank R. Fisher, superintendent of the local mine; Capt. L. V. Rice, of Robert W. Hunt & Co., engineers, of Chicago; B. J. Slining, one of the engineers; H. S. Bowen, in charge of the designing of equipment, and Van Waggoner Alling, who is a member of the National Housing Association.

After the banquet the members and guests returned to the first floor where the meeting was held. President

and other business would be postponed until the next regular meeting.

He then briefly stated that they had invited representatives of the Standard Oil company to be with them, also the other guests whose names have been mentioned above.

Mr. Fisher was first called upon for remarks, and he responded briefly, and among other things said that a number of new men would be employed in the local mine as soon as they could be secured. He concluded his remarks by inviting all of those present to visit the mine in Carlinville, and see for themselves the improvements under way.

Thomas Williamson, of Edwardsville, spoke on behalf of the Standard Oil company regarding their activities in Carlinville and vicinity. He explained that he represented Mr. Clark, who had expected to be present, but was called out of the city on important business. Mr. Williamson talked about what the Standard Oil company is going to do here by comparison with what they had done at Woodriver, one of their very largest plants. His intimation was that activities of at least as large a character would be the result of the Standard's coming to Carlinville. The company have been at Woodriver for ten years. They had no interests there until their refinery was first installed. It has been added to, and their combined interests in Woodriver amount to several million dollars. Mr. Williamson pointed out that they paid about 90 per cent of the taxes of Woodriver; the town has paved streets, electric lights, sewers—all of the modern conveniences of an up to date city. They have two school buildings, one costing $60,000, another $50,000. Their employes were all housed in good, substantial residences. He stated that the company had never objected to any of the taxes levied.In other words, they were

This was the first public utterance by any Standard Representative on what they proposed to do in Carlinville and the surrounding areas.

Captain Rice, engineer for the mines, was the next speaker. He told Commercial Club members that he came to Carlinville last May and, he himself, had put the test holes in the field in the vicinity of the town. The tests, covered the 23,000-acres that the company acquired. He had gone back to Standard and reported favorably on this field, and the activity that was taking place in Carlinville and the area was the result. He added further, that a branch railroad would be built from a point north of the C & A Railroad, straight across to the Schoper site; and that the company would install all of its own switches and lights at the mine site.

After several other speakers, Mr. Alling was called to speak. Mr. Alling was interested in the housing problem. His work was with the National Housing Association, and he had purchased property in Carlinville. He spoke about how impressed he was with Carlinville and its people and he believed that the city had a great future. He ended by saying, *"This city is now on a golden opportunity to expand."*

W.B. Otwell spoke next and stated, *"Welcome this great day for Carlinville to grow."*

After several others spoke, the meeting was adjourned and described by those in attendance as a *splendid success,* one that would be remembered for many months, if not years, to come.

With all of the new laborers arriving in Carlinville to sink the new mine shafts and build the new offices and structures at the mine sites, there suddenly became a great need for housing. The local papers ran pleas to the people of the community to open their homes to travelers. All of the town's hotels were full every night, and people were being turned away. Something needed to be done and quickly.

Carlinville Please Open Your Homes

February 17, 1918, the Carlinville Democrat issued one such plea to the citizens of Carlinville, *"It has become necessary for traveling men to telegraph ahead in order to secure accommodations at hotels in Carlinville. It has been many years since such a condition existed. It simply means that this city is entering upon a boom, which is unprecedented.*

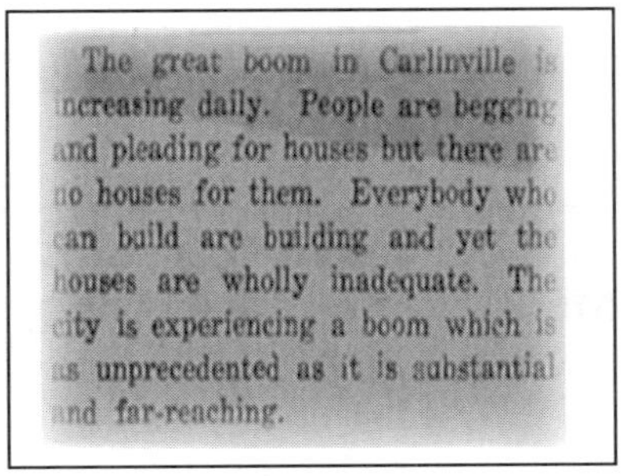

The great boom in Carlinville is increasing daily. People are begging and pleading for houses but there are no houses for them. Everybody who can build are building and yet the houses are wholly inadequate. The city is experiencing a boom which is as unprecedented as it is substantial and far-reaching.

Every night the rooms in all the hotels are occupied. In some instances guests are turned away because accommodations cannot be furnished for them.

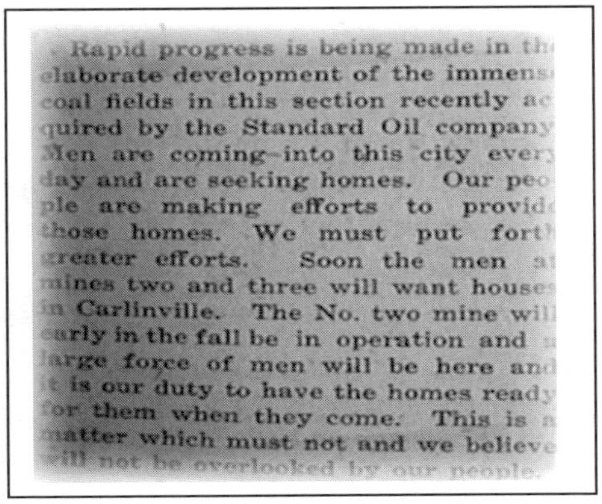

Rapid progress is being made in the elaborate development of the immense coal fields in this section recently acquired by the Standard Oil company. Men are coming into this city every day and are seeking homes. Our people are making efforts to provide those homes. We must put forth greater efforts. Soon the men at mines two and three will want houses in Carlinville. The No. two mine will early in the fall be in operation and a large force of men will be here and it is our duty to have the homes ready for them when they come. This is a matter which must not and we believe will not be overlooked by our people.

One evening last week 17 persons were unable to get rooms at one of our hotels. It is a serious matter and our citizens will necessarily have to open their homes in order to accommodate people who want to tarry in this great county seat of Macoupin, and we have confidence in our citizens to do that. There are many rooms in private residences, which could very easily be put at the disposal of the hotel people so that visitors to our city will not be embarrassed in their effort to find accommodations. If we have not awakened to the fact that it is a condition and not a theory, which confronts us, it is high time that we do so. Let everyone catch the spirit of the times, and grow in generosity and progressiveness with the demands of the occasion."

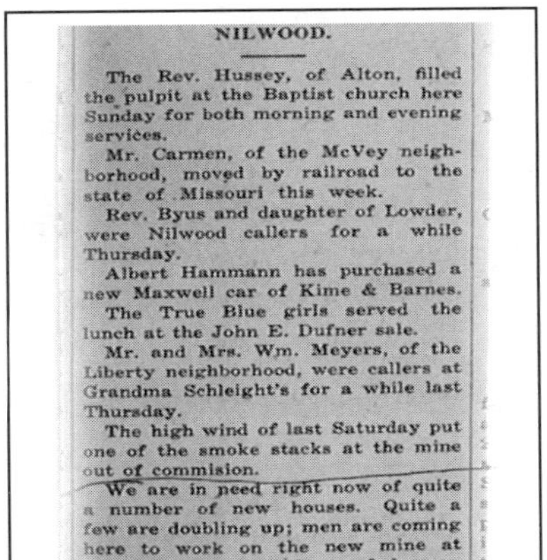

NILWOOD.

The Rev. Hussey, of Alton, filled the pulpit at the Baptist church here Sunday for both morning and evening services.

Mr. Carmen, of the McVey neighborhood, moved by railroad to the state of Missouri this week.

Rev. Byus and daughter of Lowder, were Nilwood callers for a while Thursday.

Albert Hammann has purchased a new Maxwell car of Kime & Barnes.

The True Blue girls served the lunch at the John E. Dufner sale.

Mr. and Mrs. Wm. Meyers, of the Liberty neighborhood, were callers at Grandma Schleight's for a while last Thursday.

The high wind of last Saturday put one of the smoke stacks at the mine out of commision.

We are in need right now of quite a number of new houses. Quite a few are doubling up; men are coming here to work on the new mine at Schoper. We are short of houses, barns and boarding houses. We believe it would be a good proposition for some one to build a lot of new houses.

Rev. Wunderlich and wife were called to Springfield last Thursday to the funeral of a relative.

Grandma Cullop is quite ill at this writing.

Miss Sorenson, the deaconess, held a special service at Liberty, Rural and

Nearby, the small town of Nilwood, IL was experiencing the same housing shortages. *"Nilwood needs more houses – not a vacant house in town, and many of the homes are crowded."*

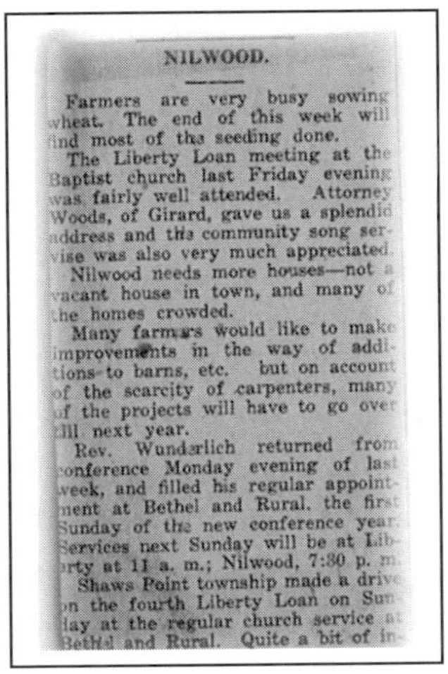

NILWOOD.

Farmers are very busy sowing wheat. The end of this week will find most of the seeding done.

The Liberty Loan meeting at the Baptist church last Friday evening was fairly well attended. Attorney Woods, of Girard, gave us a splendid address and the community song servise was also very much appreciated.

Nilwood needs more houses—not a vacant house in town, and many of the homes crowded.

Many farmers would like to make improvements in the way of additions to barns, etc. but on account of the scarcity of carpenters, many of the projects will have to go over till next year.

Rev. Wunderlich returned from conference Monday evening of last week, and filled his regular appointment at Bethel and Rural, the first Sunday of the new conference year. Services next Sunday will be at Liberty at 11 a. m.; Nilwood, 7:30 p. m.

Shaws Point township made a drive on the fourth Liberty Loan on Sunday at the regular church service at Bethel and Rural. Quite a bit of in-

Houses were being built in Carlinville, but as fast as they would go up, they were occupied and more were needed. With the mines nearing completion and the prediction that hundreds of men would be moving to the area, something needed to be done and fast. To get quality long-term workers, you had to have a place for them and their families to live. There had to be a solution to the housing problem.

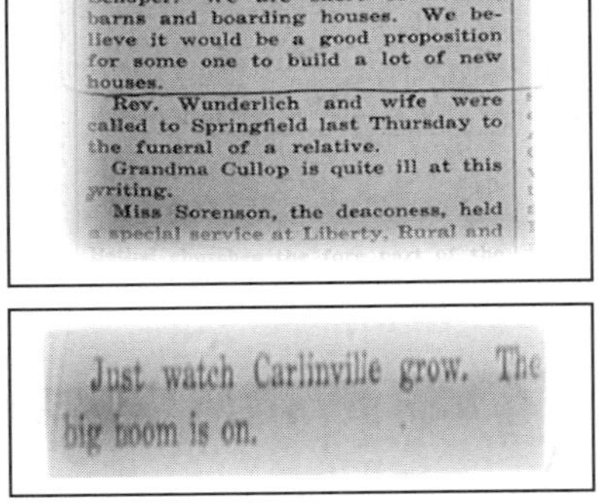

Just watch Carlinville grow. The big boom is on.

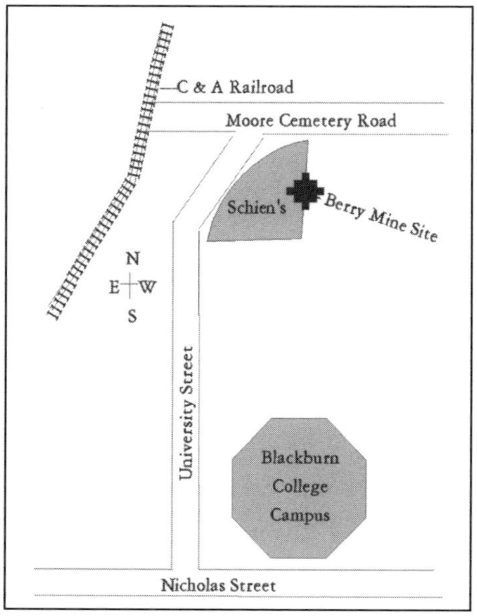

ꟾ By the first part of **1919,** the completion of mineshafts at the sites was narrowing down, but the population increase in Carlinville was not narrowing down by any means; instead, it was expanding.

☙ Berry Mine is Completed ❧

ꟾ The first shaft to be completed was dubbed "Berry Mine." It was in operation by mid **1919** and producing 2,800 tons of coal a day and employed 600 plus laborers.

ꟾ On **March 14, 1919**, the Berry Mine broke their own record for hoisting coal. The record two weeks prior was 1,216 tons of coal hoisted in 8 hours at Standard Mine No.1. This record was broken when, in 8 hours 985 boxes or 1,256 tons were hoisted. It was reported in the local paper, *"The record is gratifying to the management and shows that the mine is being worked to a high degree of efficiency."*

Note: The washhouse at the old Berry Mine site was occupied for many years by Carlinville Laundry. Inc. The old boiler and engineers building was purchased and enlarged by Schien Body and Equipment Company. Schien Body and Equipment Company still occupies that area.

*The Berry Mine in **1920**. It was located at the North end of University Street right off the C & A Railroad Line. Schein Equipment Company now occupies this site.*

☙ Schoper Mine is Completed ❧

ꟾ The second shaft was completed and went into operation in **1920**. It was named "Schoper Mine" in honor of the farmer who sold his land to the company. Schoper Mine was known as the largest coal mine in Illinois producing a total of 4,500 tons of coal a day with potential to produce even more. The local papers started running help wanted ads to hire more laborers. The completed Schoper Mine was more than 213-feet tall to the top of the smokestack. This was said to be the 2nd tallest thing in Macoupin County in those days. The tallest was the steeple on top of the 1869 courthouse. A brick and concrete building with 3-stories housed the mining offices and the base was reported

—Wanted—Laborers at Standard Oil company mine site. Apply to Schoper to superintendent.
MULVILLE BROS., Contractors.

to have been almost 30-feet deep into the ground. Schoper Lake provided the steam driven power source for the mine as well as the houses near the mine, via water through underground pipes. When all was done, Standard had invested more than $2 million in the power house, not counting another million it sank into the more than 440-feet deep mine. No wonder it was called the largest mine in Illinois.

*The completed Schoper Mine in **1920's**.*

Standard thought of everything, even a 12-bay garage on the sight for the foremen and superintendents who lived in the nearby houses. What a sight this mine must have been when in full operation.

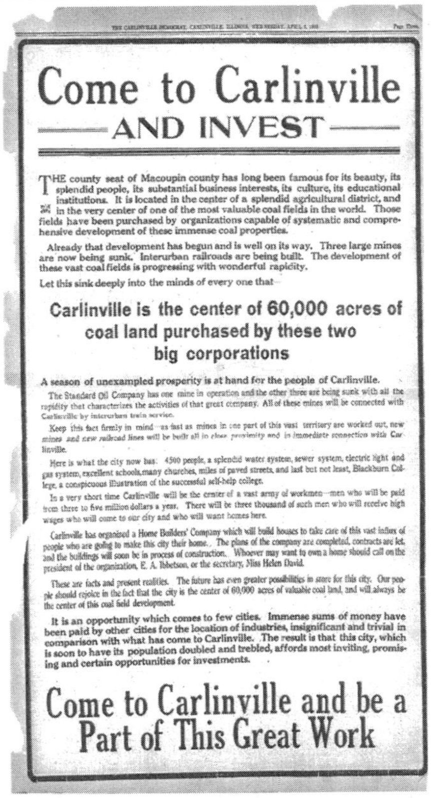

A miner's train ran 2 times a day from Carlinville to Standard City and carried the workers to Schoper Mine.

This 12-mile ride cost 10 cents.

It seemed that Carlinville was destined to become a great town with all of the extras needed to raise a mining family, except houses to live in. Carlinville was nicknamed the *"Standard Oil City."* Everyone believed that the population of Carlinville was sure to grow to 20,000 by **1920.**

An aggressive campaign had been started to entice people to come to the great city of Carlinville and invest – *"invest in the future."*

1920's *postcard view of what it looked like in the back, on any typical day at Schoper Mine, if you look closely enough you can make out the roofline of the Sears houses that were across the road.*

Northwest view at the Schoper Mine in the ***1920's.***

Above and below are photos of Schoper Power House in ***1923.***

Tipple construction at Schoper in ***1920's.***

Above and below are photos of the tipple being constructed at Schoper in ***1920's.***

Photo at the bottom of page is the northeast view of the tracks, tipple and buildings at the Schoper Mine site in the ***1920's.*** *Note the "Big Four" railroad car that carried coal from the mine.*

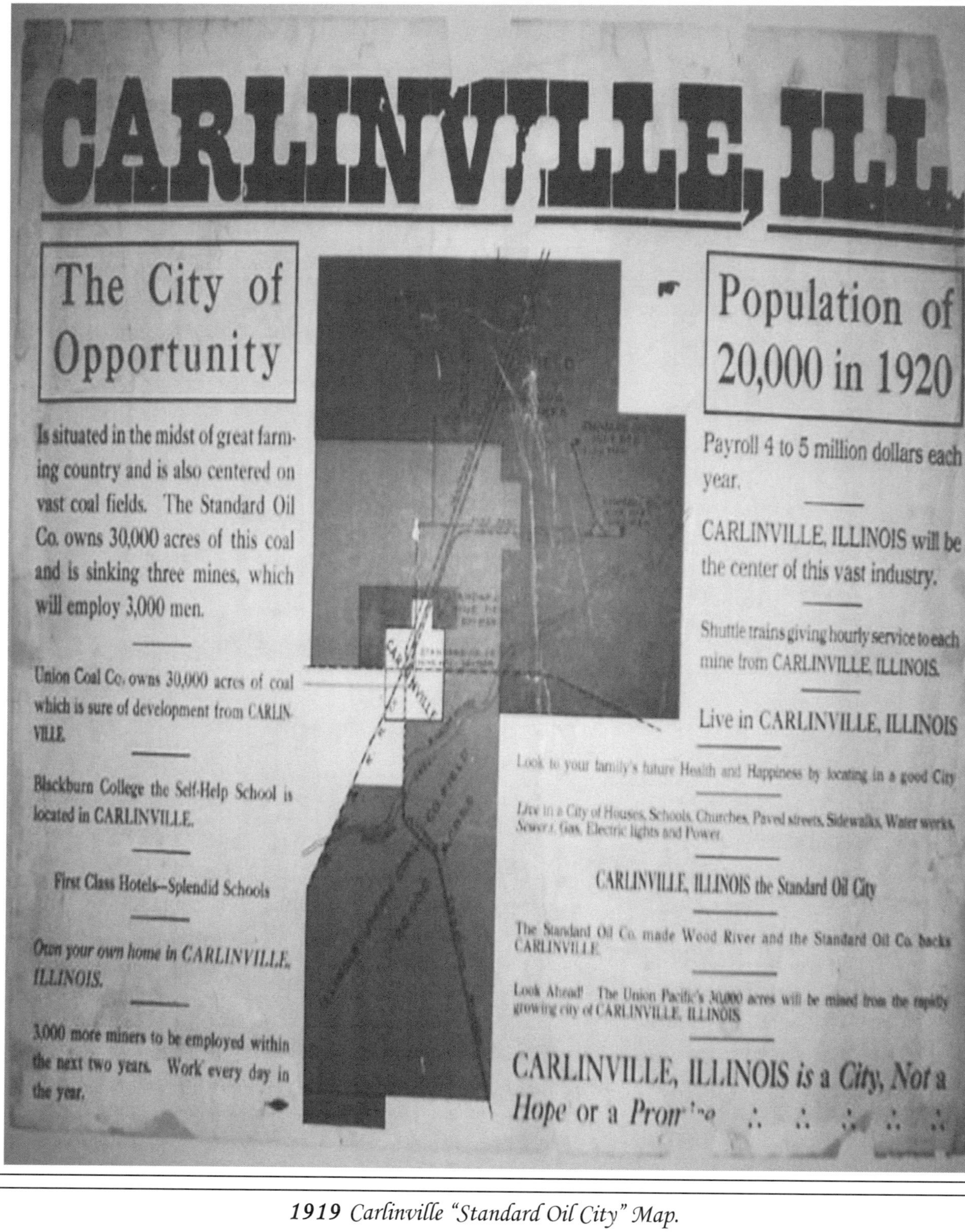

***1919** Carlinville "Standard Oil City" Map.*

Special Thank You to the Macoupin County Historical Society for a copy of the map above. The original of this hangs on the wall in the Macoupin County Historical Society Museum in Carlinville, IL.

Chapter III

"A Place for a House To Call a Home"

A Place for a House To Call a Home

☙ Standard Turns To Sears For Help ❧

❧ With the need for houses becoming so desperate, Standard turned to Sears, Roebuck & Co. for a solution to the problem. Sears could provide houses that could be built in record breaking time, but Standard needed a place to build them.

☙ Five Land Tracts Added To City Limits ❧

❧ A petition was presented on **March 20, 1918** to the Carlinville City Council in a special session, asking that a proposition for incorporating 5 tracts of land into the city limits be submitted on the **April 16, 1918**, city ballot so that the city could vote on it. On suspension of city rules, that it was too late to add a question to the ballot, the council granted the petitioner's request and the proposition was slated to go on. If the voters approved, 170-acres north of town would become part of the county seat of Macoupin. The petition called for the following tracts of land to be incorporated into the city limits: University Place consisting of 40-acres just west of Blackburn College, the Standard Oil Company's 80-acres just north of University Place, 10-acres owned by E.A. Ibbetson and Alex M. Crawford, which lay west of the Standard Oil 80, 18-acres owned by Howard Anderson north of the Ibbetson-Crawford tract, 1-acre belonging to the Bangert Estate on the hard road and 20-acres north of the Bangert property. This new addition to the city, if it became such by the voters, was bound on the west by the hard road and on the east by the road running north and west of the college known as University Street. This tract of land took in a half mile of the tracks of the C. & A. Railroad and also a half-mile of tracks owned by the Interurban line. This expansion would add $100,000 worth of taxable property to the city's budget.

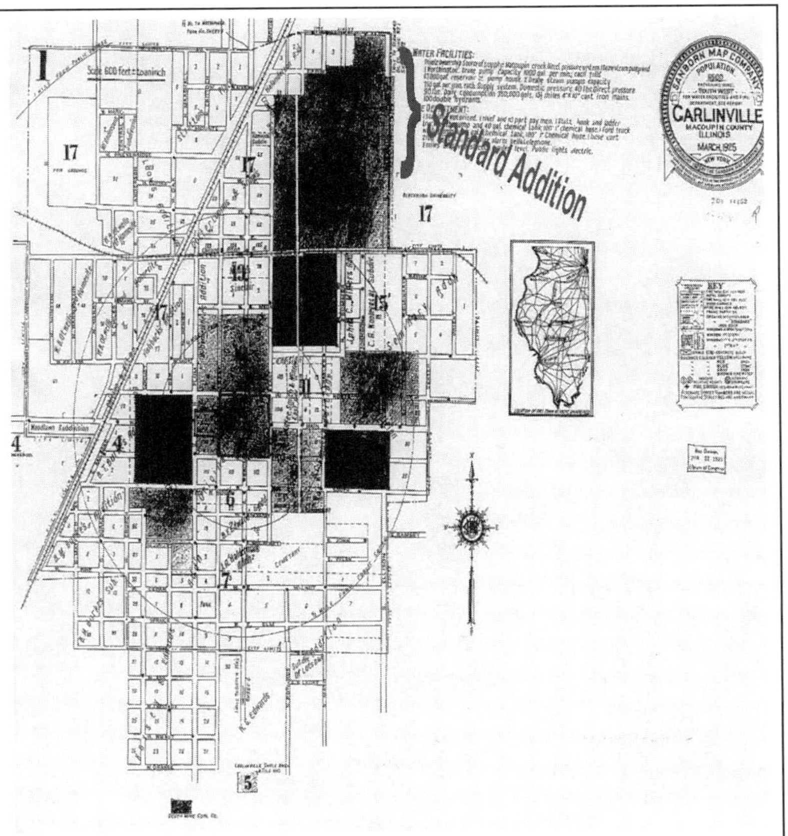

1925 Map of Carlinville, Illinois.

Note: At that time, the northern most part of the Carlinville city limits ended at Nicholas Street, which ran east to west. Some of the tracks of the Chicago & Alton Railroad actually ran along side of Nicholas Street to the college. This is the reason that Nicholas Street is so wide from the railroad crossing to University Street. Also, the hard road, was old Route 4, which goes straight north past the old Wareco Building on North Broad Street. Route 4 was rerouted later and is now the road running past Prairie Farms Dairy.

The ordinance to the right, lists the exact tracts where Carlinville's Historic Standard Addition, which contains the Sears, Roebuck & Co. Catalog Homes, is located. At the time this proposition was being voted on, this land was home to a massive wheat field. The section just south of it was called University Place, because it was directly across from the college.

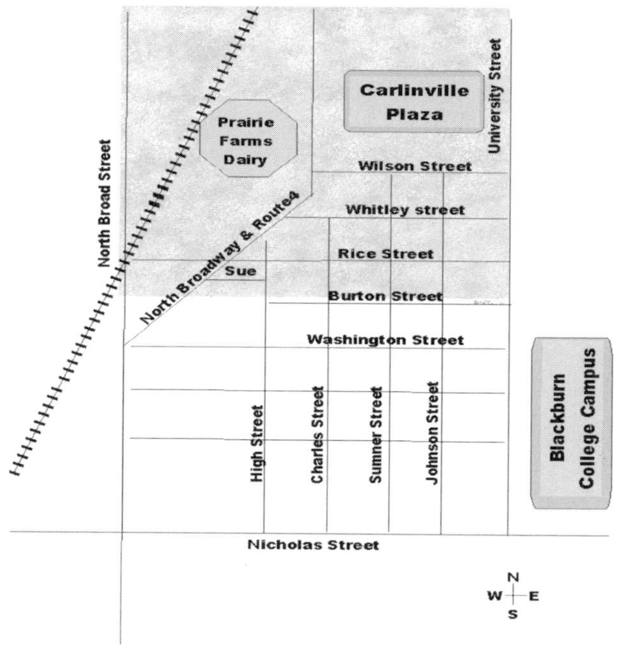

Area shaded in gray is the boundary to Standard Addition – only 9-blocks contain the Sears Catalog Homes. The 3-blocks on Whitley Street that were empty, were home to Stanolind Park and the School Buildings.

AN ORDINANCE

An ordinance amending an ordinance entitled "An ordinance in relation to the annual city election to be held on Tuesday, the sixteenth day of April, A. D. 1918, at, within and for the city of Carlinville, in the county of Macoupin and State of Illinois."

Be it ordained by the City Council of the City of Carlinville, Illinois:

Section 1. That an ordinance entitled "An ordinance in relation to the annual city election to be held on Tuesday, the sixteenth day of April, A. D. 1918, at, within and for the city of Carlinville, in the county of Macoupin and State of Illinois," adopted March 4, A. D. 1918; approved March 4, A. D. 1918, be, and the same is hereby amended so that the same shall read as follows, to-wit:

That on Tuesday, the sixteenth day of April, A. D. 1918, there shall be held at, within and for the city of Carlinville, in the county of Macoupin and State of Illinois, an election for the purpose of electing the following officers, viz; One City Marshal; two Night Policemen; one Alderman from the First ward of said city for the full term of two years; one Alderman from the Second ward of said city for the full term of two years; one Alderman from the Third ward of said city for the full term of two years and one Alderman from the Fourth ward of said city for the full term of two years.

Section 2. Also for the purpose of voting on the following questions of public policy. 1st, Shall Otwell's First Homesite addition to Carlinville, Ill., being the east part of the south half of the northeast quarter of the northeast quarter of section twenty-nine, town ten, north range seven, west of the third principal meridian, and Otwell's Second Homesite addition to Carlinville, Ill., being a subdivision of the east half of the southeast quarter of the northeast quarter of section twenty-nine and a part of the west half of the southwest quarter of the northwest quarter of section twenty-eight, town ten north range seven, west of the third principal meridian, and lot numbered three Mill Lot Subdivision, be annexed to the city of Carlinville, Ill.?

2d—Shall the West Half of the Southeast Quarter, and all of the East Half of the Southwest Quarter lying east of the center line of the "Hard Road" now extending north and south through said tract of land, and all being in section number twenty-one (21); and all the Northwest Quarter of the Northeast Quarter of section number twenty-eight (28) as heretofore platted and subdivided as "University Place" by plat recorded in Book "C" at page 42 of the records of the Recorder's office of the county of Macoupin and State of Illinois, excepting Blocks Five (5) and Twelve (12) of such subdivision, and all of said premises being in Township numbered Ten (10) North, Range numbered Seven (7) West of the Third Principal Meridian and in the county of Macoupin and State of Illinois, be annexed to the city of Carlinville, Illinois?

3d—Shall the city of Carlinville, Ill. issue bonds in the sum of three thousand five hundred dollars ($8,500.00), bearing interest at the rate of five per centum per annum, and levy a special tax for the payment of the same, for the purpose of purchasing necessary apparatus for the Carlinville Fire Department, said bonds to be issued in denominations of one hundred dollars each, and said three thousand five hundred dollars to be divided into five yearly installments of seven hundred dollars each, payable annually, beginning May 1st, A. D. 1919, together with accrued interest on the same?

Section 3. The places of voting in said city at said election shall be as follows, to-wit: In the First ward, at 124 East Main street in said city; in the Second ward in the southeast corner of the basement in the Macoupin County Court House in said city; in the Third ward in the City Hall in said city and in the Fourth ward at 214 West Main street in said city.

Section 4. The polls of said election shall be opened at the hour of seven o'clock a. m. on said day and shall continue open until the hour of five o'clock p. m. of said day.

Section 5. That the City Clerk of said city be, and he is hereby directed to publish this ordinance as an election notice, in accordance with the statute in such case made and provided.

☙ New Organization Formed In Carlinville ❧

❧ A new organization was formed called "Carlinville Home Builders Association." Its purpose was to erect houses in Carlinville in a short amount of time. The capital stock of the organization was $100,000. This stock was to be issued in shares of $25 each. The plan was for the building of houses to be paid in installments, somewhat similar to that of a Building & Loan Association. This meant that while the interest was being paid, the principal was being reduced.

❧ On **March 20, 1918**, the *Carlinville Democrat* reported more than $75,000 of Carlinville Home Builders Association stock was bought up and in short time the full amount would be purchased.

This same edition of the paper also ran an ordinance alerting would be voters to the fact that an added proposition was on the April ballot. The proposition was to incorporate several new tracts of land into the city limits of Carlinville.

❧ On **April 3, 1918,** a contract was filed in the recorder's office for the sale of a 191-acre farm tract owned by David Anderson and known as the "McGready Place." This land lay next to the Schoper tract, which was owned by Standard Oil. The sale price was $37,282.

ᘐ **April 3, 1918**, a small item in the **Carlinville Democrat** read, *"Thomas Williamson, attorney and trustee for Standard Oil, was in Carlinville Saturday looking after the interests of his clients. He is enthusiastic about Carlinville, and feels certain that a great boom is on for this city."*

ᘐ Mullville Bros., contactors for the new mine, ran a help wanted ad in the next edition of the paper, *"WANTED – Laborers at Standard Oil Mine site. Apply at Schoper to superintendent."*

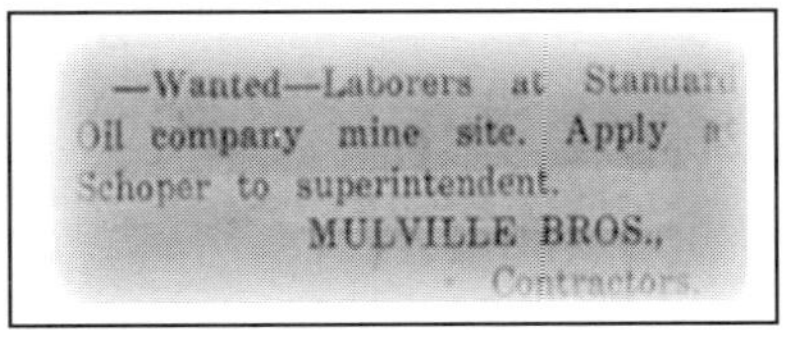
—Wanted—Laborers at Standar
Oil company mine site. Apply a
Schoper to superintendent.
MULVILLE BROS.,
Contractors.

ᘐ The **April 17, 1918**, **Carlinville Democrat**, *"Rapid progress is being made in the development of the immense coal fields recently acquired by Standard Oil. Men are coming to this city everyday and seeking homes. Our people are making efforts to provide these homes, but we must put forth greater efforts. Soon the men at mines 2 and 3 will want houses in Carlinville. The No. 2 Mine will, early in the fall, be in operation and a large force of men will be here; it is our duty to have homes ready when they come."*

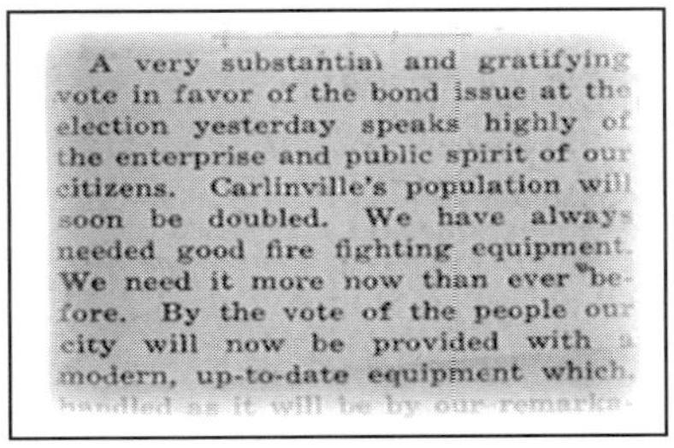
A very substantial and gratifying vote in favor of the bond issue at the election yesterday speaks highly of the enterprise and public spirit of our citizens. Carlinville's population will soon be doubled. We have always needed good fire fighting equipment. We need it more now than ever before. By the vote of the people our city will now be provided with a modern, up-to-date equipment which

This same edition also said, *"A very substantial and gratifying vote in favor of the bond issue at the election yesterday speaks highly of the spirit of our citizens. Carlinville's population will soon be doubled."* This meant that the votes were in favor and the new tracts of land would be incorporated into the city limits. At the time, no one knew what was going to happen with the new tracts of land. They assumed that Standard Oil was going to build something there, but what?

Population of 20,000 in 1920

Payroll 4 to 5 million dollars each year.

Note: Carlinville's population did grow substantially, but it never did double. People in Carlinville have heard it said that, "Carlinville's population doubled by 3,000 people overnight." That never happened. The proof of that is in newspaper articles in several places in this book. Citizens had estimated, or shall I say, dreamed, that the population of the city would reach 20,000 by 1920. That dream never became a reality – Oh, if it only had.

ᘐ **May 1, 1918**, the **Carlinville Democrat**, *"The businessmen and citizens of Carlinville know how to take advantage of the boom, which has come to the community as a result of the sinking of several mines. The housing problem is being looked at first. No town can grow without proper housing. Plans have been laid for the erection of 200 houses by Carlinville Home Builders Association and more will be built as fast as they are required."*

ᘐ The Springfield News-Record, **May 1918**, *"Carlinville is certainly to grow by leaps and bounds during the next months, and the spirit manifested by Carlinville people insures solidity."*

Please Phone Ahead

Suggestion for traveling men coming to Carlinville: Telephone ahead for hotel accommodations

ᘐ Suggestion for traveling men coming to Carlinville: Telephone ahead for accommodations.

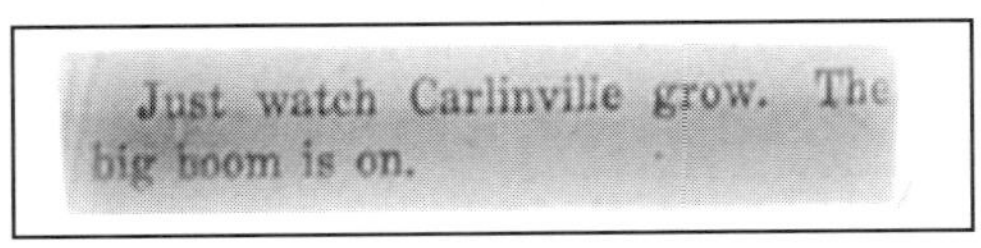
Just watch Carlinville grow. The big boom is on.

The people of Carlinville realized more & more that their city was entering upon the greatest boom in its history and the papers suggested: *"Every one should put his shoulder to the wheel and boost for a bigger and better Carlinville."*

Alexander M. Crawford and Jessie Anderson Crawford (his wife)
41 *—To—*
Thos. Williamson, Trustee.

Warranty Deed. Consideration $100.00.

Dated Jan. 10th, 1918. Recorded Jan. 11th, 1918. In Book "289" page 361. Homestead released.

Conveys the W. ½ of the S. E. ¼ of Section 21, Township 10, North, Range 7, Containing 77 acres more or less, excepting also the right of way of the Chicago and Alton Railroad Company across the same and excepting so much of said land as lies North and West of such right of way.

42 New Additions to Carlinville. One including W. ½ S. E. ¼ Sec. 21, T. 10 N., R. 7, was, on proposition for annexation, at the regular city election April 16th, 1918, carried, and by ordinance adopted by the City Council of Carlinville, Illinois, July 1st, 1918, at its regular meeting, the said additions were annexed and made a part of the City of Carlinville as certified by the Mayor and City Clerk July 2nd, 1918; plat certified by the City Engineer and ordinance recorded July 2nd, 1918, in Large Plat Book A, pages 56 and 57.

Thomas Williamson, Trustee, Mattie L., his wife
43 *—To— Warranty Deed*
Standard Oil Co., (Indiana Corporation).

Dated Feby 12th, 1918. Recorded May 18th, 1918. Book 291, page 271. Consideration $1.00.

Conveys: W. ½, S. E. ¼ Sec. 21, T. 10 N. R. 7, containing 77 acres, more of less, except coal underlying same, excepting also right of way of Chicago & Alton Railroad Company across same, and also except so much of said land as lies north and west of such right of way. Homestead Released.

Standard Oil Co., a corporation organized in State of Indiana, by W. P. Cowan, President, Attest George W. Stahl, Secy., with corporate seal attached
44 *—To— Plat*
Standard Addition to Carlinville.

Dated July 25th, 1918. Recorded August 10th, 1918. Plat Book A page 59.

Being located on south part W. ½, S. E. ¼ Sec. 21 T. 10 N. R. 7. Surveyed by S. T. Morse, Engr.

Said addition containing 12 blocks each containing 18 lots. Size of Lots in Blocks, 1, 2, 4, 5, 7, 8, is 47 ft. by 144 ft. Size of Lots in Blocks 3, 6 is 47 ft. by 141.7 ft. Lots in Blocks 9, 10 and 12 are 46.55 ft. by 144 ft. Lots in Block 11 are 46.55 ft. by 141. 7 ft.

Streets and alleys are dedicated to the use of the public forever. Coal and right to mine and remove same reserved.

This plat is made pursuant to authority given by Board of Directors of said Company.

All lines and boundaries are made with reference to a fixed and permanent monument, viz., a stone set firmly in the ground at N. W. corner of Block 4.

45 *Be it known that the Standard Oil Company, a corporation organized in the State of Indiana being the owner of all that part W. ½ S. E. ¼ Sec. 21T, 10 N., R. 7 W., 3rd P. M., Macoupin County, Ills., that lies east of the right of way of the Chicago and Alton Railroad Company* and having caused the south part of the same to be surveyed and subdivided into lots, blocks and streets by S. T. Morse, a competent surveyor and having designated such subdivision as Standard Addition to the City of Carlinville, hereby makes the attached plat of said Subd. a part hereof for the sale of lots and for all other purposes as designated by lot and block numbers on said plat, and it, the Standard Oil Company, does hereby dedicate the streets shown thereon to the public forever and it, the Standard Oil Company also dedicates a strip of land 6 ft. in width off rear end of each of said lots for use of the public forever for the sole purpose of laying and maintaining sewers, water mains, gas mains, wire conduits and overhead wire lines.

The great building boom in Carlinville is so constant and persistent that our people are becoming more or less accustomed to it. Not so with visitors. Even people riding through on the trains look out of the windows and remark about the great boom that Carlinville is experiencing. The pleasant and gratifying thing about it is that it is permanent and not fleeting.

To the left is the abstract of the land near Carlinville, which was sold to Standard Oil Company. This was dated ***February 12, 1918,*** *and is from the title search records that were done on our home.*

While Standard was working on their interests in Carlinville and the surrounding area, they were also building filling stations across the country. The photo to the right is a Standard Gas Station in the ***1920's.***

›ecial Thanks to Thomas R. Giddens, Jackson & Giddens & Judy (Giddens) White for use of the above image from their father, Paul H. Giddens' book "Standard Oil Company of Indiana - Oil Pioneer of the Midwest."

July 10, 1918, the **Carlinville Democrat**, *"Standard is adding improvements at their mine in town. They have installed a new 300 horsepower boiler and are enlarging the boiler room. A new iron stack is on the ground and will be put in place in a short time. This stack is 5-feet in diameter and is 80-feet in length. It will be erected on a 20-foot base, which makes the total height 100-feet."*

The great Standard Oil Company had come to Carlinville, Illinois for coal and they found it.

To mine the coal you need miners. They now had the land on which to build the houses they so desperately needed to lure good quality workers for their coal mines. They knew where they could get houses and build them in record time. All of the pieces were falling into place. When would the houses be ready for the miners?

Chapter IV

"*Not* By Standard, A Whole New Town"

Not By Standard, A Whole New Town

Standard City Is Born

The land where the town of Standard City sits, was purchased in **1917**, by David P. Anderson for farming. Before Mr. Anderson received the deed to the land, Standard Oil filed a suit in Macoupin County Court to stop the sale. Standard lost this suit and Mr. Anderson was allowed to purchase the land. Afterwards, Mr. Anderson offered the land to Standard, but they were no longer interested. Mr. Anderson sold the land to Mr. D.K. Rice from Gillespie, IL 3 months later. Mr. Rice laid the land out into lots to be sold. Standard Oil then sank a mine north of that land and abandoned it shortly after they sank it due to water leaking into the mine. They then sank a mine just south of Mr. Rice's land known as Schoper Mine.

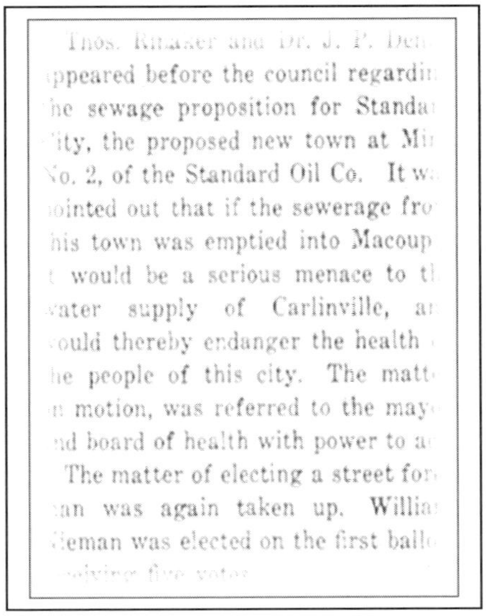
Thos. Rinaker and Dr. J. P. Den
ppeared before the council regardi
he sewage proposition for Standa
ity, the proposed new town at Mi
No. 2, of the Standard Oil Co. It w
ointed out that if the sewerage fro
his town was emptied into Macoup
would be a serious menace to t
vater supply of Carlinville, a
ould thereby endanger the health
he people of this city. The matt
n motion, was referred to the may
nd board of health with power to a
The matter of electing a street for
an was again taken up. Willia
Neman was elected on the first ball

Note: People have believed for many years that Standard Oil owned this land and they had it platted out into town lots to sell. This is not so. Standard never owned the land where the town of Standard City sits. They tried to bar the sale of it so that they could buy it, but they were unsuccessful. It was Mr. Rice from Gillespie who owned the land when it was platted out into lots for houses and it was he who named the town Standard City. The name Standard City was chosen because of Standard, but Standard had nothing to do with the town's platting or building.

May 15, 1918, Carlinville City Council news, *"Thomas Rinaker and Dr. J.P. Denby appeared before the council regarding the sewage proposition for Standard City, the proposed new town at Mine No. 2 of Standard Oil. It was pointed out that if sewage from this town emptied into Macoupin Creek it would be a serious menace to the water supply of Carlinville, and endanger the health of the people of this city. The matter, on motion, was referred to the mayor and board with the power to act upon it."*

Note: Growing up in this area, we've always heard it said Standard City was a mining town built by Standard Oil for workers to live in. With the name Standard City, who wouldn't think it? It may have been named after Standard Oil, but they didn't build it. Many people have said that Schoper was the name of this town first and then it was changed to Standard City; this was certainly not the case. It was the town of Standard City from the very beginning. Standard City and Schoper were always two completely separate entities connected by mining.

In **1918**, Standard City was booming just like Carlinville. Houses and businesses were constructed and at one time, there were 12 business buildings in Standard City. They included: a garage, 2 hardware stores, a shoe store, a grocery store, a meat market, a dance hall, a theatre, a lumberyard, a bakery, a school and a dry goods store. The grocery store was owned by Tom Danelvich and John Bianchetti and was in business until the **1960's**. In its early days, the town was home to a lot of illegal alcohol makers.

Standard City, Illinois, became incorporated in **1920**; and although it never really grew much in size from what it was in the mining years, it is still there today and home to about 100 people who take pride in living there. Part of Standard City is located in Nilwood Township and the other part is in Shaw's Point Township. The 1st school at Standard City was located in a bunk house owned by Standard Oil.

SALE
of SCHOOL HOUSES and SITES
and CERTAIN PERSONAL PROPERTY

ATWATER SCHOOL
SATURDAY, JUNE 2, '56

OAKLAND SCHOOL
SATURDAY, JUNE 2, '56

RURAL SCHOOL
SATURDAY, JUNE 2, '56

CHESTERFIELD SCHOOL
SATURDAY, JUNE 2, '56

Read Legal Notice to Left

WARREN MOFFETT,

MIKE MAKUH,

Standard City Schoolhouse built in ***1924.*** *This building was erected when Standard City School, Robley School and Lind Grove School consolidated because of financial hardships. There were two classrooms both upstairs and downstairs in this building. The old wooden school, that was used for many years, was moved into Standard City and added to the grocery store. This school became part of the Carlinville School District in* ***1947.***
In ***1956*** *this building, as well as several others in the Carlinville District, was sold at auction. The auction sale bill is to the left.*

Note: People have said that all of the Sears Homes from Standard City and Schoper were moved to Standard Addition. This is untrue. They did move many of the 12 homes from Schoper to other locations, but none were moved to Standard Addition. The last home moved from Schoper is located on Route 108. There were 2 homes remaining at Schoper when this last home was moved. Of those remaining 2 homes, 1 fell into such a dilapidated state that it was torn down. The other was home to the mine superintendent until he left in the 1940's after the mine was sold. The mine operated for a few years with barely 100 employees and then closed again. It would go through 2 more owners through the years only to be reclaimed by the Abandoned Mine Land Department of Natural Resources and demolished on August 27, 2002. Unfortunately, the last house remained vacant until it was destroyed by fire in the fall of 1996. This house was a Langston Model. (See Chapter X – Standard Oil Company Unloads, for a complete history of the last house moved from Schoper.)

Don't you wish you could still buy a house for $513.00?

The house pictured on the top of the next page, stands in the middle of Standard City today and is proof of the fact that there are still Sears, Roebuck and Co. Homes there. The owner of this home probably does not realize that he owns an original Sears, Roebuck & Co., Catalog Home called the Rosita Model. The Rosita Model is one of those rare Sears, Roebuck & Company Modern Homes that only appeared in the catalog for a short time and then vanished. This house model is hard to find.

Note: The Rosita Model was offered for sale for only 4-years 1917-1921. So I think that it is safe to assume that this house was probably built for the miners also.

*Below is the advertisement for the house as it appeared in the **1914** Sears, Roebuck & Co. Honor Bilt Modern Homes Catalog. Only 300 of these homes were sold in the 4-years it was offered by Sears.*

Are There More Sears Homes In Standard City?

Could there be more Sears Homes still located in Standard City? The answer is **YES.** There could be Honor Bilt, Simplex Ready Homes or even Standard Built Homes? There are many that certainly do look familiar and have characteristics of Sears Modern Homes.

Could the house in the photo to the left be a Sears Home? It is located in Standard City.

How about the homes in the photos below, they might be Sears Homes? These homes are also located in Standard City.

The advertisement to the left, appeared in the **May 19, 1918** issue of the **Carlinville Democrat**. It enticed people to come to Standard City and see *"The Center of Activity"* of the new mine town being built. *"Buy a lot for $20 down and $10 a month."*

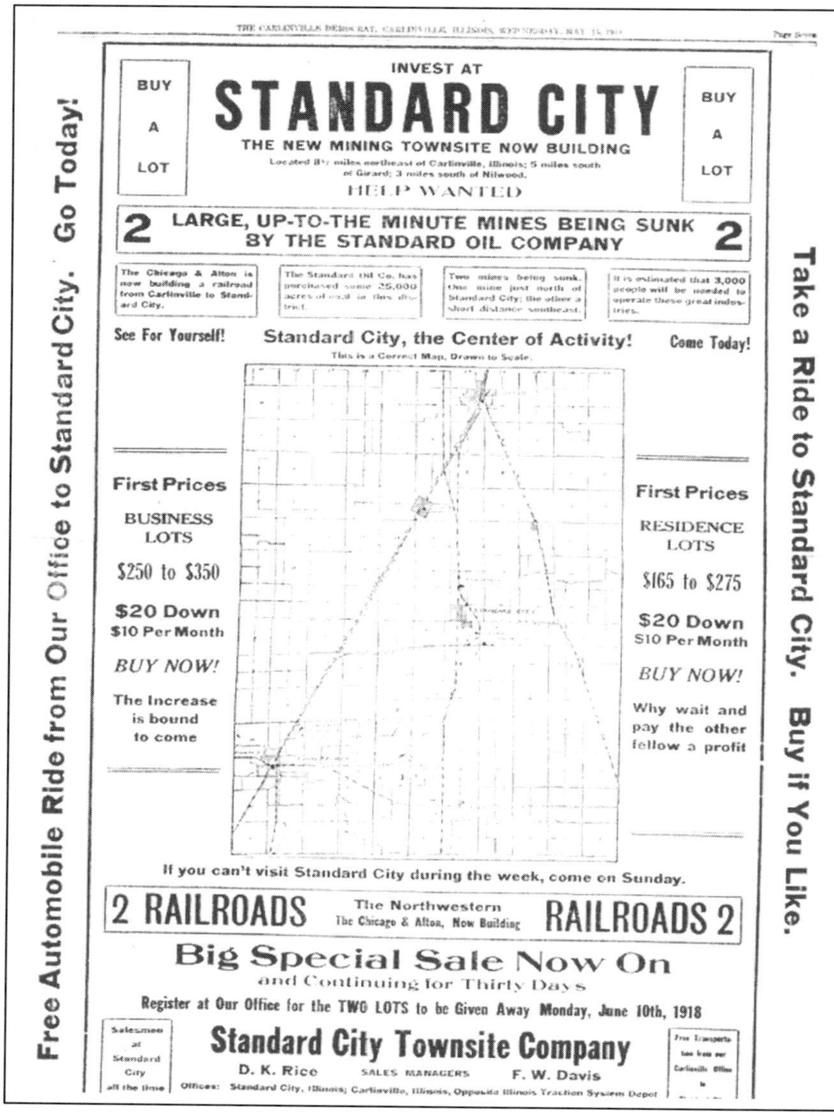

Free Automobile Ride from Our Office to Standard City. Go Today!

INVEST AT

BUY A LOT

STANDARD CITY

BUY A LOT

THE NEW MINING TOWNSITE NOW BUILDING

Located 8½ miles northeast of Carlinville, Illinois; 5 miles south of Girard; 3 miles south of Nilwood.

HELP WANTED

2 LARGE, UP-TO-THE MINUTE MINES BEING SUNK BY THE STANDARD OIL COMPANY 2

The Chicago & Alton is now building a railroad from Carlinville to Standard City.

The Standard Oil Co. has purchased some 25,000 acres of coal in this district.

Two mines being sunk. One mine just north of Standard City; the other a short distance southeast.

It is estimated that 3,000 people will be needed to operate these great industries.

See For Yourself! **Standard City, the Center of Activity!** Come Today!

This is a Correct Map, Drawn to Scale.

First Prices

BUSINESS LOTS

$250 to $350

$20 Down
$10 Per Month

BUY NOW!

The Increase is bound to come

First Prices

RESIDENCE LOTS

$165 to $275

$20 Down
$10 Per Month

BUY NOW!

Why wait and pay the other fellow a profit

If you can't visit Standard City during the week, come on Sunday.

2 RAILROADS The Northwestern / The Chicago & Alton, Now Building **RAILROADS 2**

Big Special Sale Now On

and Continuing for Thirty Days

Register at Our Office for the TWO LOTS to be Given Away Monday, June 10th, 1918

Salesmen at Standard City all the time

Standard City Townsite Company

D. K. Rice SALES MANAGERS F. W. Davis

Offices: Standard City, Illinois; Carlinville, Illinois, Opposite Illinois Traction System Depot

Take a Ride to Standard City. Buy if You Like.

*The sign below, is "standing proof" that Standard City has been there since **1920**.*

There are many old and interesting small buildings in Standard City today, which were built during coal mining days. Some of these were used as offices; some could have even been used as a 1-room home for a miner. There is no doubt they date back to the **1920's**, so they have some connection to the mines.

The small stone block building in the photo to the left still stands in Standard City today as a monument to the coal mining days.

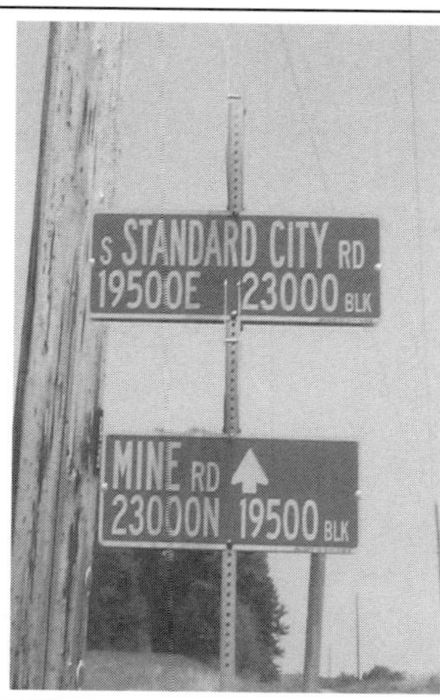

Chapter V

"The Houses Are Coming, The Houses Are Coming"

The Houses Are Coming, The Houses Are Coming!

☙ Standard To Build Houses in Carlinville ❧

❧ Standard Oil had now secured the land on which they would erect the much-needed houses for the mining families. Now all they needed were the houses.

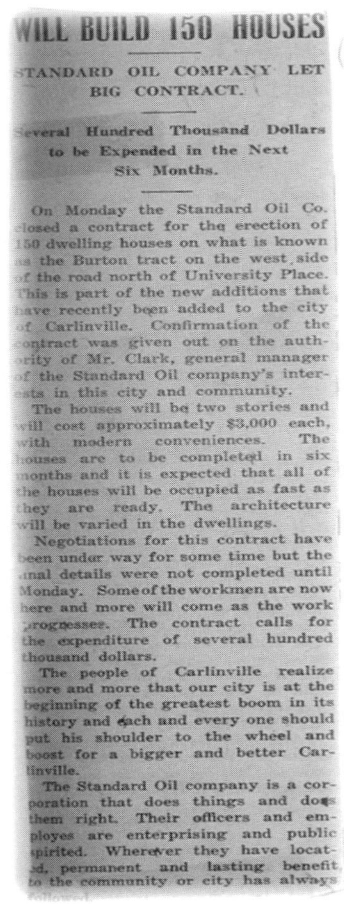

WILL BUILD 150 HOUSES

STANDARD OIL COMPANY LET BIG CONTRACT.

Several Hundred Thousand Dollars to be Expended in the Next Six Months.

On Monday the Standard Oil Co. closed a contract for the erection of 150 dwelling houses on what is known as the Burton tract on the west side of the road north of University Place. This is part of the new additions that have recently been added to the city of Carlinville. Confirmation of the contract was given out on the authority of Mr. Clark, general manager of the Standard Oil company's interests in this city and community.

The houses will be two stories and will cost approximately $3,000 each, with modern conveniences. The houses are to be completed in six months and it is expected that all of the houses will be occupied as fast as they are ready. The architecture will be varied in the dwellings.

Negotiations for this contract have been under way for some time but the final details were not completed until Monday. Some of the workmen are now here and more will come as the work progresses. The contract calls for the expenditure of several hundred thousand dollars.

The people of Carlinville realize more and more that our city is at the beginning of the greatest boom in its history and each and every one should put his shoulder to the wheel and boost for a bigger and better Carlinville.

The Standard Oil company is a corporation that does things and does them right. Their officers and employes are enterprising and public spirited. Wherever they have located, permanent and lasting benefit to the community or city has always

☙ Contract Let For 150 Dwelling Houses ❧

❧ **Carlinville Democrat, June 12, 1918**, *"On Monday Standard Oil closed a contract for building 150 dwelling houses on the Burton tract on the west side of the road north of University Place."* ***Note: This was part of the new tracts of land that had been added to the city limits of Carlinville.***

Although the article stated there were to be 150 homes built in Carlinville's Standard Addition, there were actually 156 erected. Since facilities for housing permanent employees were poor and insufficient in Carlinville at the time, Standard contracted with Sears, Roebuck & Co. to buy a large number of their Honor Bilt pre-cut houses.

"Confirmation of the contract was given out on authority of Mr. Clark – general manager of Standard Oil's interests in this city and this community." The houses were ordered from the Sears, Roebuck & Co. lumberyard in Cairo, IL. When completed, they were to be sold to the mining employees at prices ranging from $3,370 to $4,060 on an installment plan with 10-years to pay them off. Average payments were $30 to $40 per month.

Note: It has been said through the years that this was a million-dollar order for Sears, Roebuck & Co. Unfortunately, Sears did not keep many sales records regarding their homes, so this could not be verified. Although I am sure that this is the largest order that Sears, Roebuck & Co. had ever received for the houses, I do not think that it was a million-dollar order – the math does not add up. Even if you use the highest price of house style ordered, $4060.00, and multiply it by 192, which was the actual order placed with them by Standard Oil, that only comes to $779,520.00; and if you then tack on maybe $25,000.00 for shipping, it still doesn't equal a million-dollars. I am more inclined to believe that this was a million-dollar housing project for Standard Oil. They had to purchase the land and hire the laborers to build the homes. What I am positively, beyond any doubt sure of, is that it was the largest single ticket order for homes that Sears, Roebuck & Co. ever received.

❧ The **Carlinville Democrat, June 9, 1918** *"Our people are very much interested in the new mines that Standard Oil is sinking on the Schoper land northeast of Carlinville. The company has authorized the announcement that when the two mines are in full operation they will hoist 1,200 cars of coal a week. To get this amount of coal out they will require probably 1,000 or more men. The company is now actively engaged in developing the property. Undoubtedly the company will purchase many things they need in this city. The money that the company spends here will help our town. The money that their employees spend here, and will spend here, is and* *will help this city. The population of Carlinville will, beyond any question, be largely increased in the next few years. Standard Oil will bring people here. They have already expended a large amount of money in Carlinville and this is only the beginning."*

ൽ We Need A Road ൿ

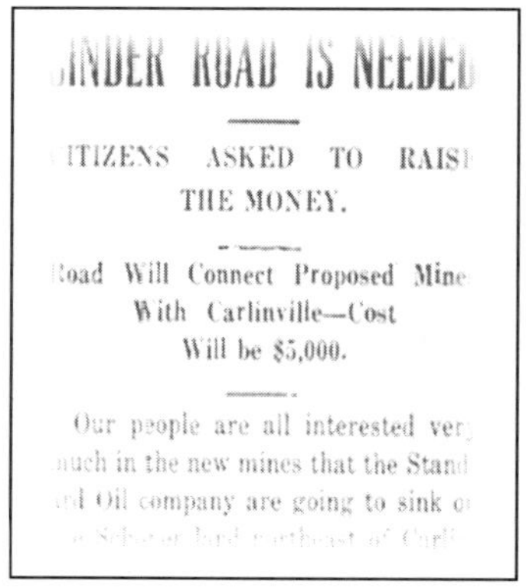
INDER ROAD IS NEEDE

ITIZENS ASKED TO RAIS THE MONEY.

oad Will Connect Proposed Mine With Carlinville—Cost Will be $5,000.

Our people are all interested ver nuch in the new mines that the Stan d Oil company are going to sink o

ൿ The **June 9, 1918** paper urged Carlinville citizens to raise money for a new cinder road to connect the two new mine sites. This was a much-needed improvement to the road as the existing roads were dirt and at times of heavy rain, not passable. If they were going to build houses at Schoper then a cinder road was most definitely needed.

When Standard Oil decided on the houses to erect in Carlinville, they chose houses that were priced in the middle of the scale of what Sears had to offer through their pre-cut Honor Bilt Modern Homes program. They were not the smallest and definitely not the largest homes, but they were considered "mansions" in size by miners who were used to living in 1-room, a tent, or a shack. The houses that were built in Standard Addition included 9 different styles. These styles were ordered from the Sears, Roebuck & Co. Honor Bilt Modern Homes Catalog. 8 of these styles were easily identified as the Langston, Madelia, Roseberry, Lebanon, Warrenton, Whitehall, Gladstone and Carlin. One style to this day is still unidentified. I have dubbed this house "The Unknown Model."

*A Cinder Road Near Carlinville in **1920's**.*

The much needed cinder road is today, simply called "Cinder Road."

*Late **1800's** – early **1900's** Standard Oil Station.*

Standard knew what houses they were going to build and where they were going to build them. Now all they needed were the builders and the houses themselves.

ൽ

Chapter V – 1

"History of Sears, Roebuck & Co. Modern Homes"

History of Sears, Roebuck & Co. Honor Bilt Modern Homes Program

Get It All From Sears – Even A Home

Sears, Roebuck & Co. started their Modern Homes Department in late **1895** when they began selling building materials from their mail order catalog. The department was not very successful until a Sears, Roebuck & Co. employee took it upon himself to turn it into a moneymaker. He launched a very aggressive advertising campaign and had a dream of selling houses through the mail.

The 1st catalog devoted entirely to mail-order homes was issued in **1908**. It featured 22 styles in prices ranging from $600 to $2,300. That one price tag got you the plans, specifications and building materials right down to the nails and doorknobs.

In that **1908** catalog, any American with a building site and $725.00 could order an entire 2-story, 6-room home through the mail. This 6-room home found on the pages of that catalog could be delivered anywhere in the country via the railroad.

Left is a photo of the 1908 Sears, Roebuck & Co. Modern Homes Catalog Cover

It came pre-cut and complete in every detail from the front porch to the trim. The pre-cut homes had each part numbered to help ease the would-be carpenter in building them. After the customer placed his order from the special "Modern Homes" catalog, the pieces of his new home were sent out by rail with the delivery arranged so that each shipment arrived at exactly the right time in the production schedule. The lumber for the floor and framing came first, already notched and mitered, again, with each piece numbered for position. Then came the rest. These came complete with a set of building plans bound in a black leather instruction manual that bore the customers name in gold. The 75-page instruction manual told step-by-step how to build and assemble the home right down to painting and varnishing. The manual also carried a warning, ***"Do not take anyone's advice as to how your building should be erected."***

The aggressive marketing campaign, enticing renters and would-be first-time homeowners to buy, made the idea of the mail-order home extremely desirable.

According to Sears, Roebuck & Co. records, the first mail-order house was sold in **1909**.

After that, orders came and came, so much so that Sears, Roebuck & Co. purchased a lumber mill in Mansfield, LA in **1909** to meet the demands for orders of their pre-cut homes.

ꝏ In **1911** they purchased a lumberyard in Cairo, IL, in **1912** a millwork plant in Norwood, OH.

ꝏ In **1918** Sears began offering credit for their houses, payable in monthly installments with a 6% interest rate. In the years to come, they would eventually loan money for everything including the lot on which to build the home.

Mr. R. W. Sears at his desk.

Although most of its Business was with individuals, Sears also dealt directly with corporations and organizations. They also sold houses to companies for towns near their factories. By **1926,** more than 37,000 families nationwide had built a Sears, Roebuck & Co. Modern Home.

The 40 plus acre Cairo, IL lumberyard made it possible for Sears, Roebuck & Co. to pre-cut houses and store them in large quantity amounts waiting for the potential orders to come – and they did come.

ꝏ By **1926,** Sears was selling homes to fit every budget, from a modest $525.00 cottage to a mansion for $5,000.00 called the Magnolia, which is seen on the **1918** catalog cover.

ꝏ **1929** saw more than 49,000 of their pre-cut homes already built and would be the highest point in the sale of homes for Sears with a volume of over $12 million in sales and nearly half that amount in mortgages.

ꝏ Country buyers could even buy a matching outhouse for $23.00 to $41.00. The outhouse came pre-cut as well, with complete assembly instructions.

ꝏ If a person wanted to build his own subdivision, Sears could fill that order as well.

ꝏ When business began to falter in the early **1930's**, Sears began carrying mortgages on both the material and the labor for the homes it sold, but the economy had slowed even with the accommodating terms Sears offered. People just simply couldn't afford to buy a home. The Depression forced the Modern Homes Department deeper and deeper into the red.

ꝏ In **1935**, Sears limited their mail-order homes business to just home sales with no financing.

By 1936, the Sears Mail Order Homes Department would print their last catalog. The homes went on to sell for a couple of years, using the same catalog. The mail-order house-building program was never revived. In all, Sears claimed they sold more than 100,000 of their homes across the country.

Sears sold other homes in addition to the Honor Bilt Homes program, including Standard Built Homes, which were of a lesser quality than the Honor Bilt, Simplex Sectional Cottages and even Simplex Sectional Garages and barns.

*Right is the **1918** Sears, Roebuck & Co. Honor Bilt Modern Homes Catalog Cover.*

Honor Bilt vs. Standard Built

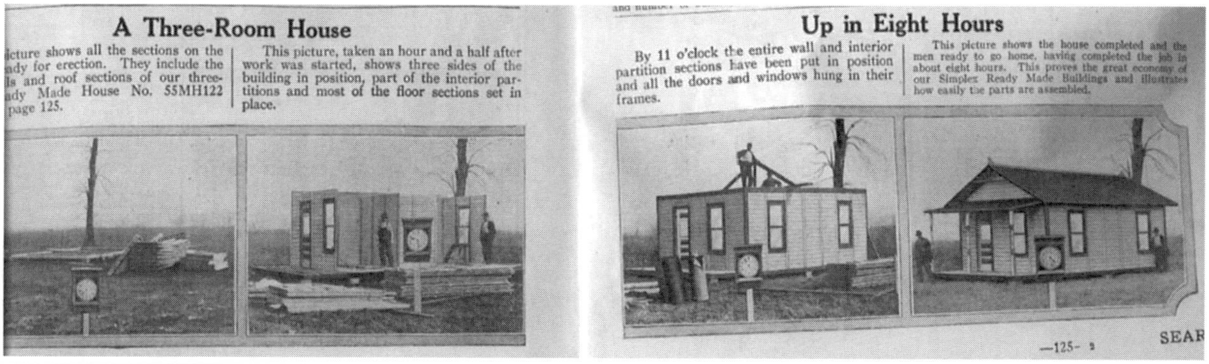

Sears even had a home that you could build in a day – just think, you could receive the kit in the morning and sleep in it that night.

*When the Modern Homes Department first offered homes for sale, Sears used model numbers for each house design; but in **1914** each house design was given a name in addition to the number. House designs that the department discontinued before **1914** never received a name.*

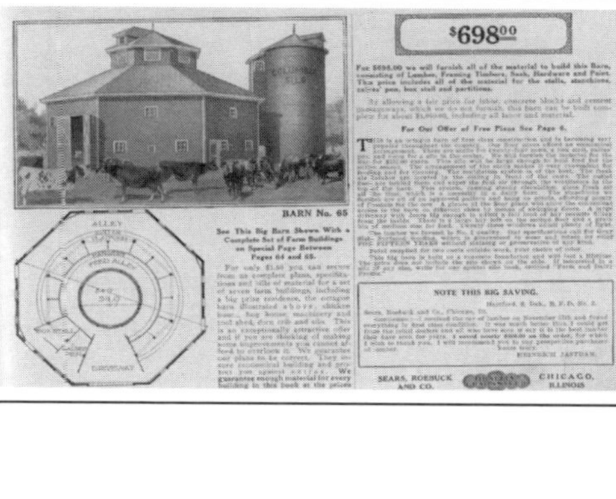

Throughout the 40-year span that Sears sold their pre-cut homes, they offered more than 600 designs. Some home styles were not marked and the company prided themselves on modifying and adapting floor plans to have ***"Every Customer Satisfied."*** So you could live in a Sears Home and not even know it.

ᘛ The postwar housing boom, boosted sales from 125 houses per month in **1920** to 250 plus houses a month in **1929** and at one particular point 324 houses in one month from the Cairo, IL plant alone.

ᘛ By **1942** the sale of their homes ceased and the Sears, Roebuck & Co. Modern Homes Department was dissolved.

Left is the last Sears, Roebuck & Co. Modern Homes Catalog Cover.

Sears offered other catalogs in addition to their Honor Bilt Homes, such as Building Materials and Millwork, Modern Plumbing, Modern Heating, Modern Lighting, and Modern Furnishings. All were designed to make things easy and affordable for the homeowner.

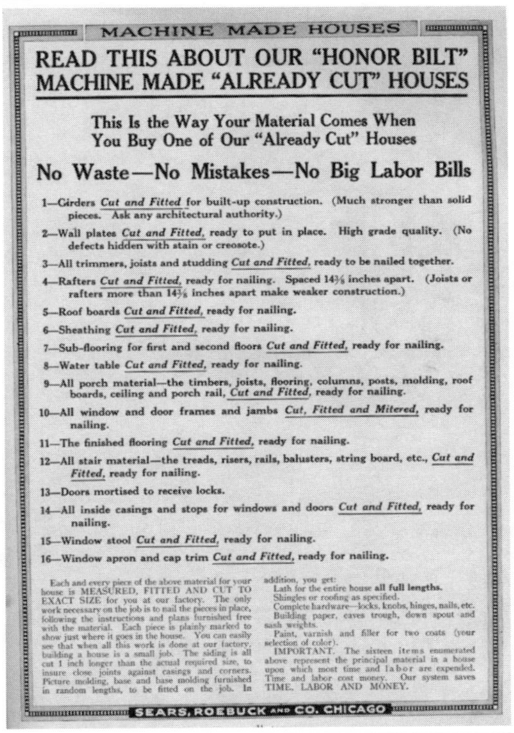

MACHINE MADE HOUSES

READ THIS ABOUT OUR "HONOR BILT" MACHINE MADE "ALREADY CUT" HOUSES

This Is the Way Your Material Comes When You Buy One of Our "Already Cut" Houses

No Waste—No Mistakes—No Big Labor Bills

1—Girders *Cut and Fitted* for built-up construction. (Much stronger than solid pieces. Ask any architectural authority.)

2—Wall plates *Cut and Fitted,* ready to put in place. High grade quality. (No defects hidden with stain or creosote.)

3—All trimmers, joists and studding *Cut and Fitted,* ready to be nailed together.

4—Rafters *Cut and Fitted,* ready for nailing. Spaced 14⅜ inches apart. (Joists or rafters more than 14⅜ inches apart make weaker construction.)

5—Roof boards *Cut and Fitted,* ready for nailing.

6—Sheathing *Cut and Fitted,* ready for nailing.

7—Sub-flooring for first and second floors *Cut and Fitted,* ready for nailing.

8—Water table *Cut and Fitted,* ready for nailing.

9—All porch material—the timbers, joists, flooring, columns, posts, molding, roof boards, ceiling and porch rail, *Cut and Fitted,* ready for nailing.

10—All window and door frames and jambs *Cut, Fitted and Mitered,* ready for nailing.

11—The finished flooring *Cut and Fitted,* ready for nailing.

12—All stair material—the treads, risers, rails, balusters, string board, etc., *Cut and Fitted,* ready for nailing.

13—Doors mortised to receive locks.

14—All inside casings and stops for windows and doors *Cut and Fitted,* ready for nailing.

15—Window stool *Cut and Fitted,* ready for nailing.

16—Window apron and cap trim *Cut and Fitted,* ready for nailing.

Each and every piece of the above material for your house is MEASURED, FITTED AND CUT TO EXACT SIZE for you at our factory. The only work necessary on the job is to nail the pieces in place, following the instructions and plans furnished free with the material. Each piece is plainly marked to show just where it goes in the house. You can easily see that when all this work is done at our factory, building a house is a small job. The siding is all cut 1 inch longer than the actual required size, to insure close joints against casings and corners. Picture molding, base and base molding furnished in random lengths, to be fitted on the job. In addition, you get:

Lath for the entire house **all full lengths.**
Shingles or roofing as specified.
Complete hardware—locks, knobs, hinges, nails, etc.
Building paper, eaves trough, down spout and sash weights.
Paint, varnish and filler for two coats (your selection of color).

IMPORTANT. The sixteen items enumerated above represent the principal material in a house upon which most time and labor are expended. Time and labor cost money. Our system saves TIME, LABOR AND MONEY.

SEARS, ROEBUCK AND CO. CHICAGO

MACHINERY AND TOOL HOUSE No. 219

16'0" — 36'0"
MACHINERY AND TOOL SHED
FLOOR PLAN

$138.00

For $138.00 we will furnish all the material to build this Machinery and Tool House, consisting of all the Lumber, Siding, Roofing, Sash, Hardware and Paint.

By allowing a fair price for labor this machinery and tool house can be built complete for about $185.00.

WE WILL furnish the plans, specifications and bill of material for $1.00. This amount will be allowed when you send us your order for the material. For a SPECIAL OFFER by which you can secure complete plans for five other outbuildings and a big residence for only $1.50, see special page between pages 64 and 65.

Up to date farmers everywhere recognize the advisability of having a separate building for machinery and tools. Much money is wasted by not taking proper care of the machines required on the average farm throughout the year. The rain rusts them quickly and otherwise impairs their usefulness. When protected in a building of this kind they are always in good shape. The building will also make a handy workshop for necessary repairs on all kinds of machinery, vehicles, harness and other odd jobs.

The machinery and tool house above illustrated is 16 feet wide by 36 feet long and is 13 feet 6 inches high from the foundation to the peak. It is lighted with ten windows, five on each side. It has two large swinging doors at each end and is roofed with our 3½-Ply Best-of-all Roofing, **guaranteed for twelve years' wear.**

MILK HOUSE No. 224

$101.00

For $101.00 we will furnish all the material to build this Milk House, consisting of Lumber, Siding, Flooring, Shingles, Hardware, Door, Sash and Paint.

By allowing a fair price for labor, this milk house can be built complete for about $150.00.

WE WILL furnish the plans, specifications and bill of material for $1.00. This amount will be allowed when you send us your order for the material. For a SPECIAL OFFER by which you can secure complete plans for five other outbuildings and a big residence for only $1.50, see special page between pages 64 and 65.

MILK TROUGH
SHELF
MILK ROOM 11'0"x13'0"
FLOOR PLAN

The milk house above illustrated is a solidly constructed building, 13 feet wide by 11 feet long and is 16 feet high from foundation to peak. It has two milk troughs in which the milk cans can be set to cool until they are ready to be taken to the creamery or until the milk is disposed of otherwise. On the left hand side a shelf runs from the door to the side wall and extends along the side wall to the milk trough in the rear.

To the farmer who wants to keep his milk in a sanitary condition, this building at the price which we charge for all the materials is a first class proposition. We guarantee a good grade of material all the way through and agree to furnish a sufficient quantity to build it according to our plans.

SEARS, ROEBUCK AND CO. QUALITY GUARANTEED CHICAGO, ILLINOIS

—121—

Honor Bilt Homes Factory

THESE are reproductions of actual photographs taken for the purpose of giving our cu tomers a better idea of the tremendous stocks carried at our lumber mills and millwo factories IN READINESS TO SHIP AT ONCE. Here you will find a larger stock building material than will be found carried by any retail dealer.

ABOVE—MODERN HOMES DIVISION

As shown in pictures, our millwork is carefully stored in dampproof, dirtproof buildings, assuring our customers of receiving only bright, clean stock. Here you will find about 90,000 Doors, over 100,000 Windows, millions of feet of hard and soft wood Moldings and Columns, Newels, Stair and Porch Material; also Flooring in equally large quantities.

ABOVE

A birds-eye view of our 17-Acre Sash and Door Factory in Ohio.

LEFT—DOOR WAREHOUSE

ABOVE—ORDER PRICING DIVISION

This book of Modern Homes shows a great variety of homes suitable fo every section of the country: East, west, north and south. All our design have proved successful. We guarantee our plans to be correct. Selec your house direct from this book and send us your order today! Otherwis write us for full information about the house you are interested in. Ou architects are at your service.

ABOVE—BUILT-UP COLUMN WAREHOUSE

Pages 17, 18 and 19 portray just a glimpse of our architectural department facilities. We exercise every care in originating house plans for all requirements.

Facilities and Offices

EVERY day hundreds of orders for materials and requests for plans and estimates come from all parts of the country. Efficient workers give every kind of service. We have the largest organization of its kind in the world. Our architects, draftsmen and expert women advisers have mastered home building problems.

ABOVE—MODERN HOMES DIVISION

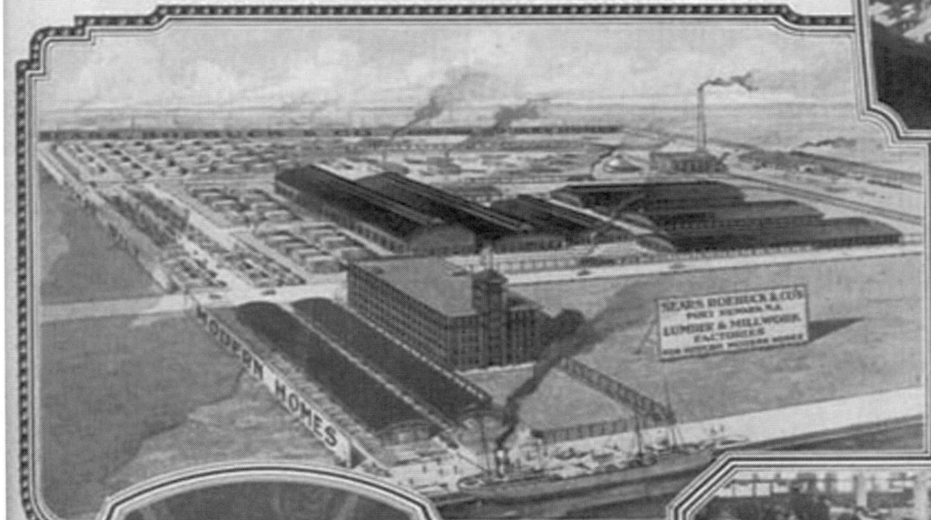

LEFT—A birdseye view of our huge Newark, N. J., Lumber and Millwork Plant. It covers a total of 40 acres, connecting with railroad terminals and ocean going freighters. The plant consists of fourteen manufacturing buildings and has the latest and most efficient modern machinery. Good service is assured, because the plant has a capacity of 100,000,000 feet of lumber per annum.

LEFT—MOLDING WAREHOUSE

RIGHT—MODERN HOMES DIVISION

These modern facilities and mammoth stocks make it possible for us to furnish you EXTRA GOOD QUALITY and EXTRA QUICK SERVICE, while our prices enable you to make A SUBSTANTIAL SAVING.

BELOW—A BIRDSEYE VIEW OF OUR BIG 40-ACRE LUMBER PLANT IN ILLINOIS

ABOVE—WINDOW WAREHOUSE

If you plan erecting a house not described in this book, send us a full list of the material you need and get our free estimate. We are in a position to quote low prices on lumber, millwork, roofing, hardware, plumbing goods, heating plant, electric lighting fixtures, paint and wall paper in any quantity.

SEARS, ROEBUCK AND CO. Page 45

Modern Plumbing & Heating Systems

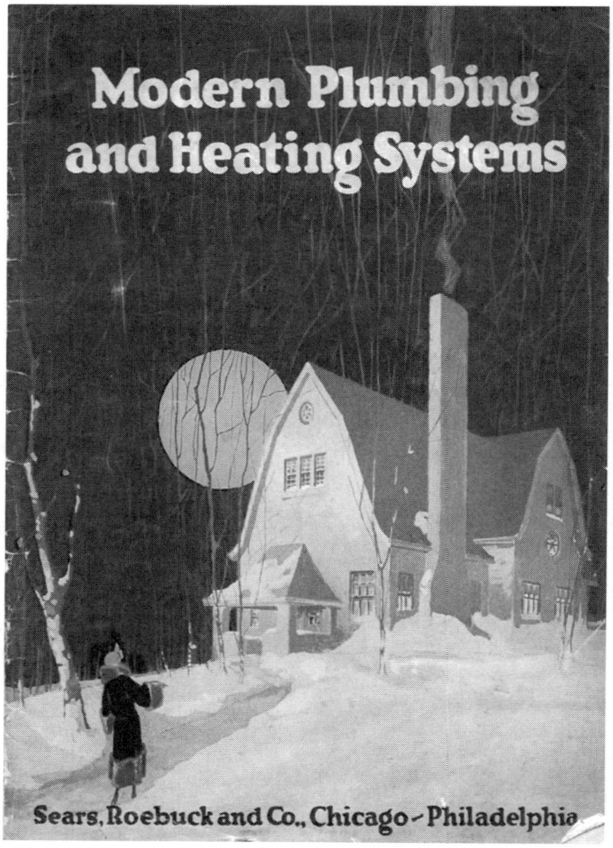

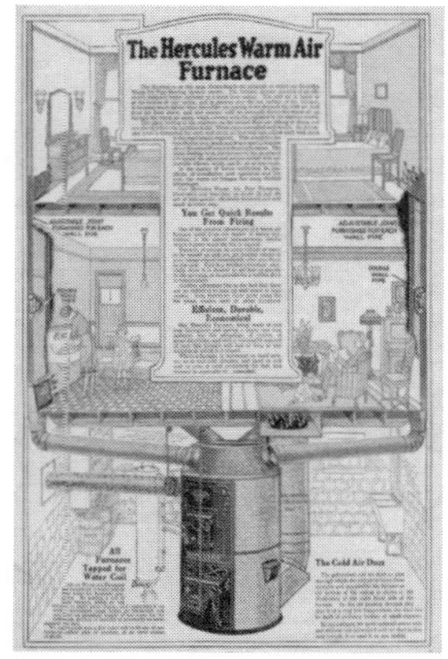

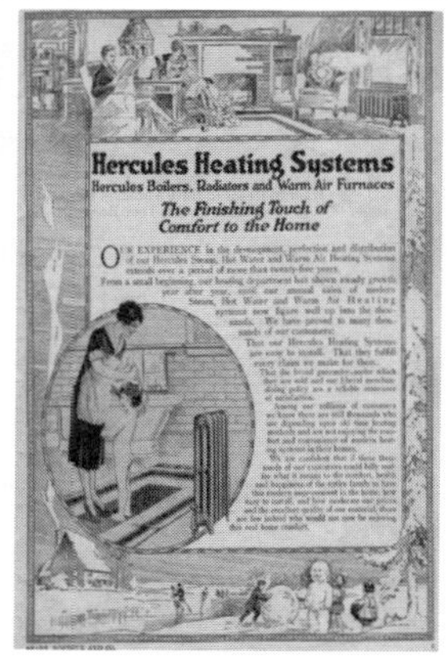

*Photos above are from **1927** "Modern Plumbing and Heating Systems" catalog by Sears, Roebuck & Co.*

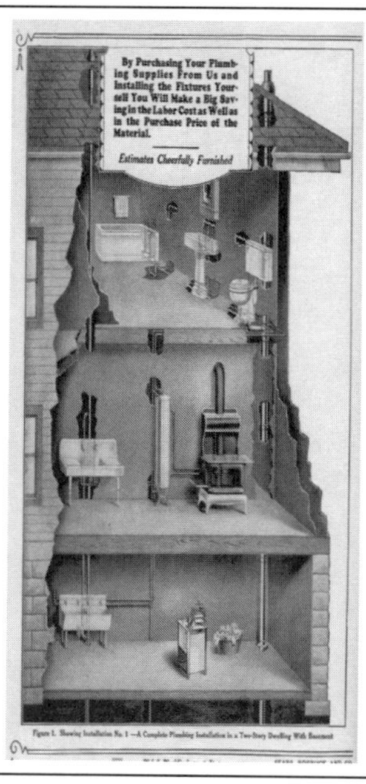

*Pictures above and right are from the **1928** "Instructions for Installing Modern Plumbing Systems" catalog by Sears, Roebuck & Co.*

Building Materials

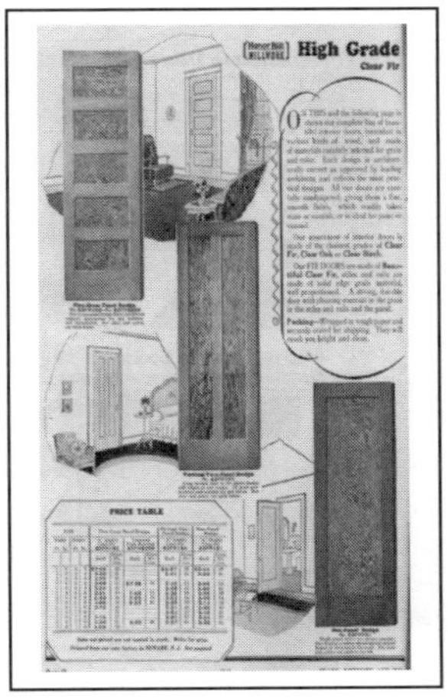

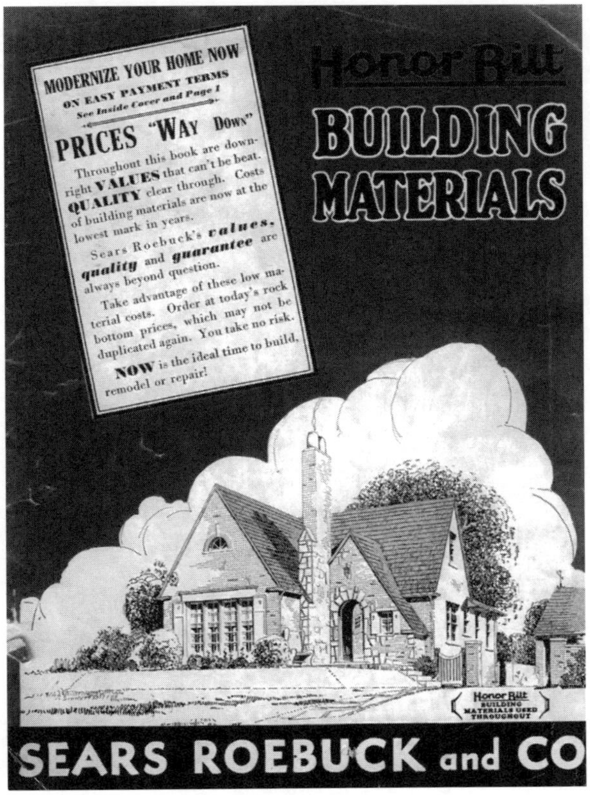

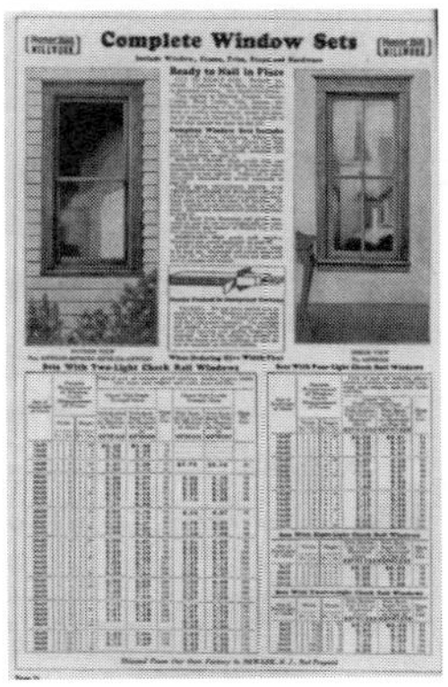

Photos above are from the ***1931*** *"Honor Bilt Building Materials" catalog by Sears, Roebuck & Co.*

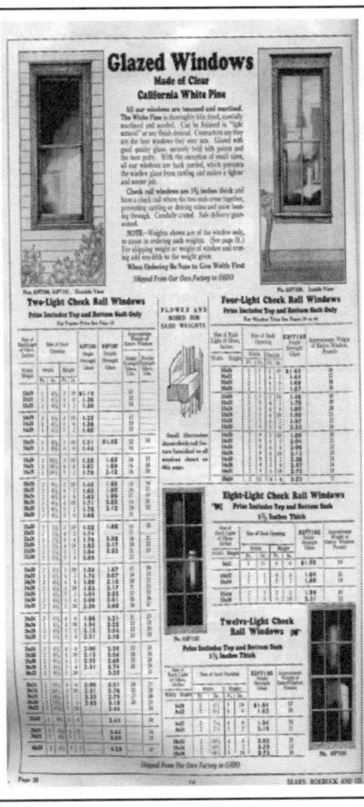

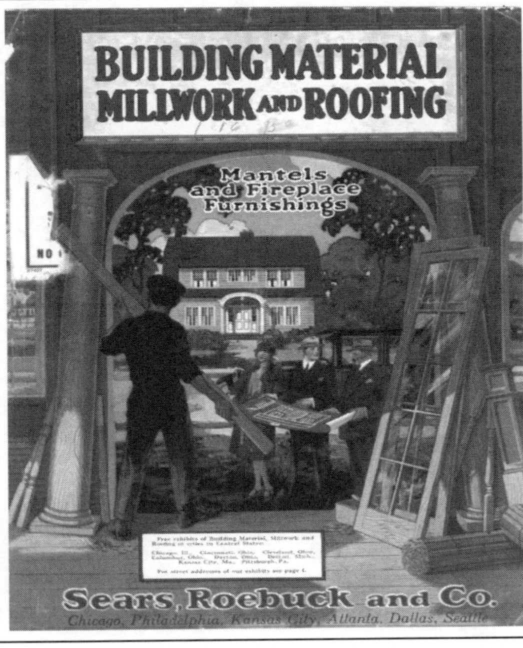

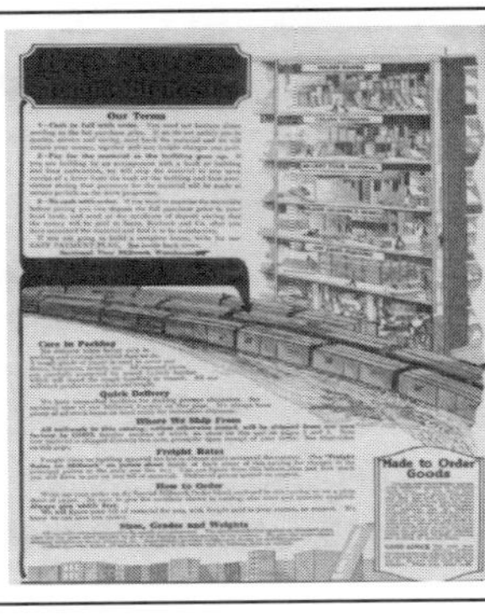

Photos above, to the right are all from the ***1927*** *"Building Material, Millwork and Roofing" catalog by Sears, Roebuck & Co.*

Electric Light and Power for Country Homes

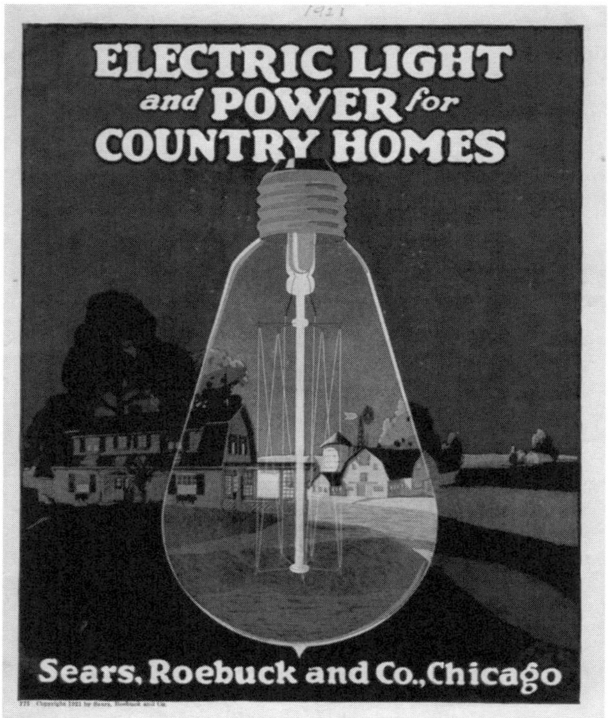

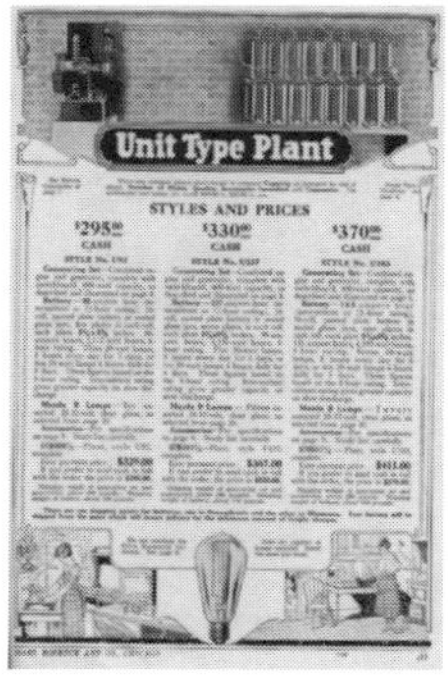

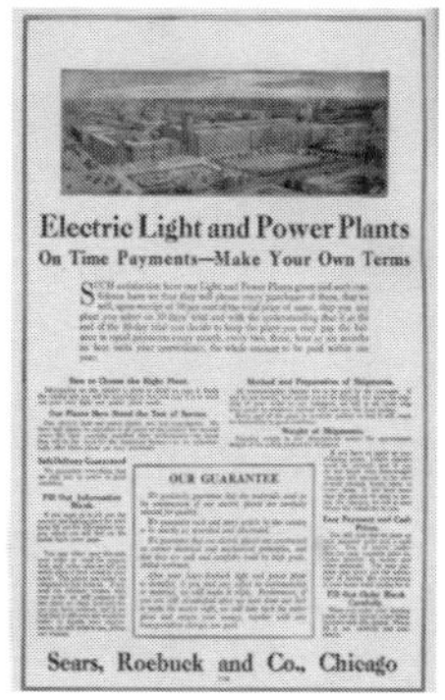

Photos above are from the ***1921*** *"Electric Light & Power for Country Homes" catalog by Sears, Roebuck & Co.*

Photos above are from the ***1928*** *"Homart Fashions In Lighting" catalog by Sears, Roebuck & Co.*

Chapter VI

"A Neighborhood *Bilt* With Honor"

A Neighborhood *Bilt* With Honor

The Houses Arrived

A temporary railroad spur was laid into Standard Addition and the houses came via boxcars on the C & A Railroad. This temporary spur was laid along High Street and then down Rice Street. The houses were then let off on the spur and manually maneuvered, or pulled by horses or mules, to their building site.

Each Standard Addition house came in 2 boxcars. The 1st boxcar carried the basics needed to raise the house. The 2nd boxcar carried everything needed to finish the house. Each house kit also included two trees for the front yard.

Laying of Temporary Railroad Spur.

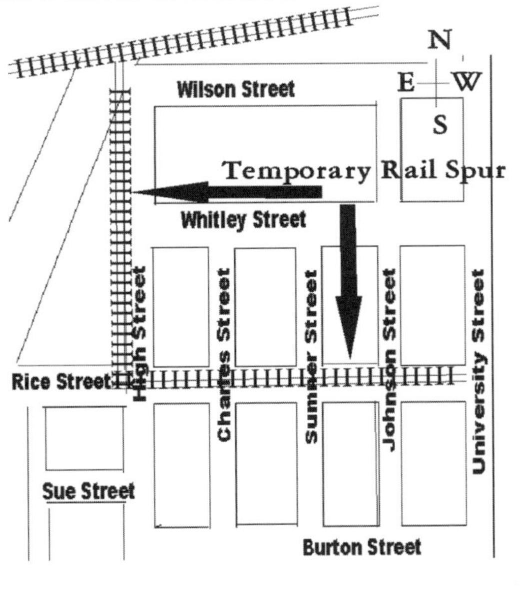

Larger houses sold by Sears were sent in more boxcars. The largest house that Sears sold, the Magnolia, took 7 boxcars to ship.

The building of the 1st homes in Standard Addition was delayed due to harvesting of the wheat crop that occupied the area. Because of the scarcity of wheat and flour, the government would not allow the wheat crop to be destroyed; they had to wait for it to grow and be harvested before they could build. It was said that, as the 1st houses were going up, they were actually threshing the wheat from the field right beside them.

*Right is wheat threshing in **1918**.*

*Left is an Illinois farmer at work on the wheat harvest in his field in **1918**.*

☙ First Sears Houses Are Built ❧

❧ In the late fall of **1918,** the building of the houses got underway with the employment of local laborers to do the carpentry. Standard Oil hired a very tough female foreperson to oversee the building. Her name was Elizabeth Spaulding, and she had a reputation for hiring men in the morning and firing them by noon if she didn't think that they were performing to her expectations. She patrolled the 9-block home building area on horseback throughout the day. One would never know when she would happen by. This was said to have kept workers on their toes. It is also said that her horse died near the end of her reign as foreperson and that it is buried somewhere in Standard Addition.

1918 photo of the C & A Railroad Tracks just north of Carlinville where the houses were brought by rail.

Note: Some have said that it was Sears who built the Standard Addition houses and not Standard Oil. If this were true, why did Standard hire a foreperson to oversee the building and how could she be the one who hired and fired the laborers if in fact it was Sears who was in charge of the building? It doesn't add up, does it? I believe that it was Standard Oil who was in charge of the building of the homes and that they themselves secured the builders. Also, the letter from Standard Oil to Sears praising them for the quality of the houses changed. The 1st few times that it appeared in the Modern Homes catalogs it said, "The 200 houses which you built…" The letter was then changed to say, "The 200 houses which you supplied…" This is proof, to me, that Standard Oil built the homes in Carlinville, Schoper and Wood River, not Sears. Whoever dictated that first letter was not very well informed.

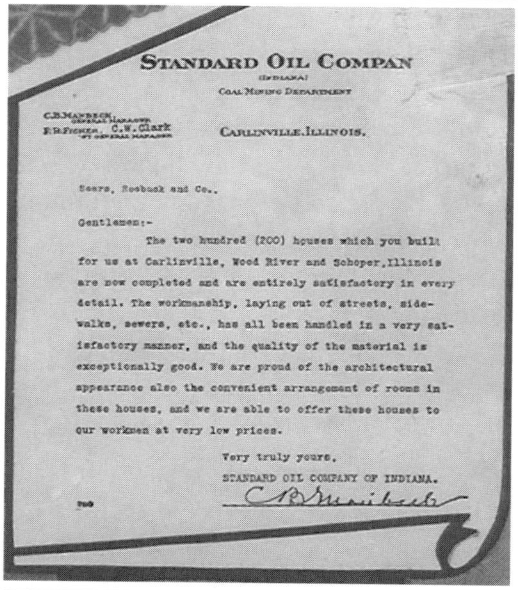

STANDARD OIL COMPAN[Y]
(INDIANA)
COAL MINING DEPARTMENT

C.B. MANBECK, GENERAL MANAGER
E.R. FISHER, C.W. Clark, ASST GENERAL MANAGER

CARLINVILLE, ILLINOIS.

Sears, Roebuck and Co.,

Gentlemen:-

The two hundred (200) houses which you built for us at Carlinville, Wood River and Schoper, Illinois are now completed and are entirely satisfactory in every detail. The workmanship, laying out of streets, sidewalks, sewers, etc., has all been handled in a very satisfactory manner, and the quality of the material is exceptionally good. We are proud of the architectural appearance also the convenient arrangement of rooms in these houses, and we are able to offer these houses to our workmen at very low prices.

Very truly yours,
STANDARD OIL COMPANY OF INDIANA.
C.B. Manbeck

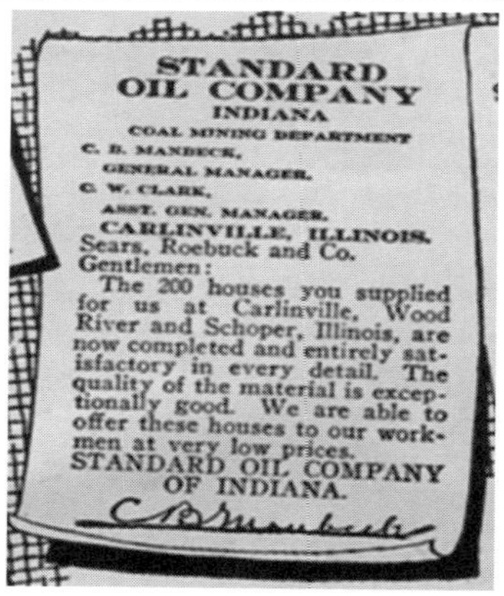

STANDARD OIL COMPANY
INDIANA
COAL MINING DEPARTMENT
C. B. MANBECK,
GENERAL MANAGER.
C. W. CLARK,
ASST. GEN. MANAGER.
CARLINVILLE, ILLINOIS.
Sears, Roebuck and Co.
Gentlemen:
The 200 houses you supplied for us at Carlinville, Wood River and Schoper, Illinois, are now completed and entirely satisfactory in every detail. The quality of the material is exceptionally good. We are able to offer these houses to our workmen at very low prices.
STANDARD OIL COMPANY OF INDIANA.
C.B. Manbeck

A section of Standard Addition was divided into 12-blocks and platted out into lots. The Sears homes were built in 9-blocks of those 12. Each block contained 18 lots – 9 on each side of the street. The lots were laid out with boundaries using a permanent and fixed monument, which was a stone firmly set in the ground at the northwest corner of block number 4. **Note: Block 4 is the 1100 block of High Street. I have searched many times and have never found a stone that could be a permanent marker.**

Lots in blocks 1, 2, 4, 5, 7, & 8 are each 47-feet by 144-feet. Lots in blocks 3 & 6 are each 47-feet by 141.7-feet. Lots in blocks 9, 10 & 12 are each 46.55-feet by 144-feet and lots in block 11 are 46.55-feet by 141.7-feet. All of these lots are at a slight angle – northeast to southwest – as the part of the Chicago & Alton Railroad line that was used when platting out the area was also at this slight slant.

Note: I get so angry when people say Standard Addition is a 12-block area on the north edge of Carlinville. Standard Addition is the entire northwest edge of Carlinville, from Burton Street, which is just south of the Sears homes, all of the way north to the railroad tracks and from University Street West all of the way to North Broad Street. This includes Rice Street from North Broadway Street to North Broad Street, the Carlinville Plaza, Prairie Farms Dairy and even the Chinese Restaurant. Only 12-blocks of this section were platted into lots for houses. The reason that only 9-blocks contain Sears Homes is because the 1100 block of Johnson Street on the east side was home to the Standard Addition portable school buildings. Directly to the west of them was home to Stanolind Park, which took up both the east and west sides of the 1100 block of Sumner and Charles Streets, as well as the west side of the 1100 block of High Street. This area is now home to a trailer park, which is not connected with the Sears Homes in Standard Addition. So let it be known that Standard Addition is home to a 9-block area that houses the largest collection of Sears, Roebuck & Co. Catalog Homes in the world. It is definitely NOT a 12-block area. A trailer is NOT a Sears Catalog Home. Sorry I had to unload – I feel better now.

Right is the original plat description from the abstract to our home.

Standard Oil Co., a corporation organized in State of Indiana, by W. P. Cowan, President, Attest George W. Stahl, Secy., with corporate seal attached
44 *—To— Plat Standard Addition to Carlinville.*

Dated July 25th, 1918. Recorded August 10th, 1918. Plat Book A page 59.

Being located on south part W. ½, S. E. ¼ Sec. 21 T. 10 N. R. 7. Surveyed by S. T. Morse, Engr.

Said addition containing 12 blocks each containing 18 lots. Size of Lots in Blocks, 1, 2, 4, 5, 7, 8, is 47 ft. by 144 ft. Size of Lots in Blocks 3, 6 is 47 ft. by 141.7 ft. Lots in Blocks 9, 10 and 12 are 46.55 ft. by 144 ft. Lots in Block 11 are 46.55 ft. by 141. 7 ft.

Streets and alleys are dedicated to the use of the public forever. Coal and right to mine and remove same reserved.

This plat is made pursuant to authority given by Board of Directors of said Company.

All lines and boundaries are made with reference to a fixed and permanent monument, viz., a stone set firmly in the ground at N. W. corner of Block 4.

Be it known that the Standard Oil Company, a corporation organized in the State of Indiana being the owner of all that part W. ½ S. E. ¼ Sec. 21T, 10 N., R. 7 W., 3rd P. M., Macoupin
45 County, Ills., that lies east of the right of way of the Chicago and Alton Railroad Company and having caused the south part of the same to be surveyed and subdivided into lots, blocks and streets by S. T. Morse, a competent surveyor and having designated such subdivision as Standard Addition to the City of Carlinville, hereby makes the attached plat of said Subd. a part hereof for the sale of lots and for all other purposes as designated by lot and block numbers on said plat, and it, the Standard Oil Company, does hereby dedicate the streets shown thereon to the public forever and it, the Standard Oil Company also dedicates a strip of land 6 ft. in width off rear end of each of said lots for use of the public forever for the sole purpose of laying and maintaining sewers, water mains, gas mains, wire conduits and overhead wire lines.

Below is Standard Addition construction ***1918 – 1919.***

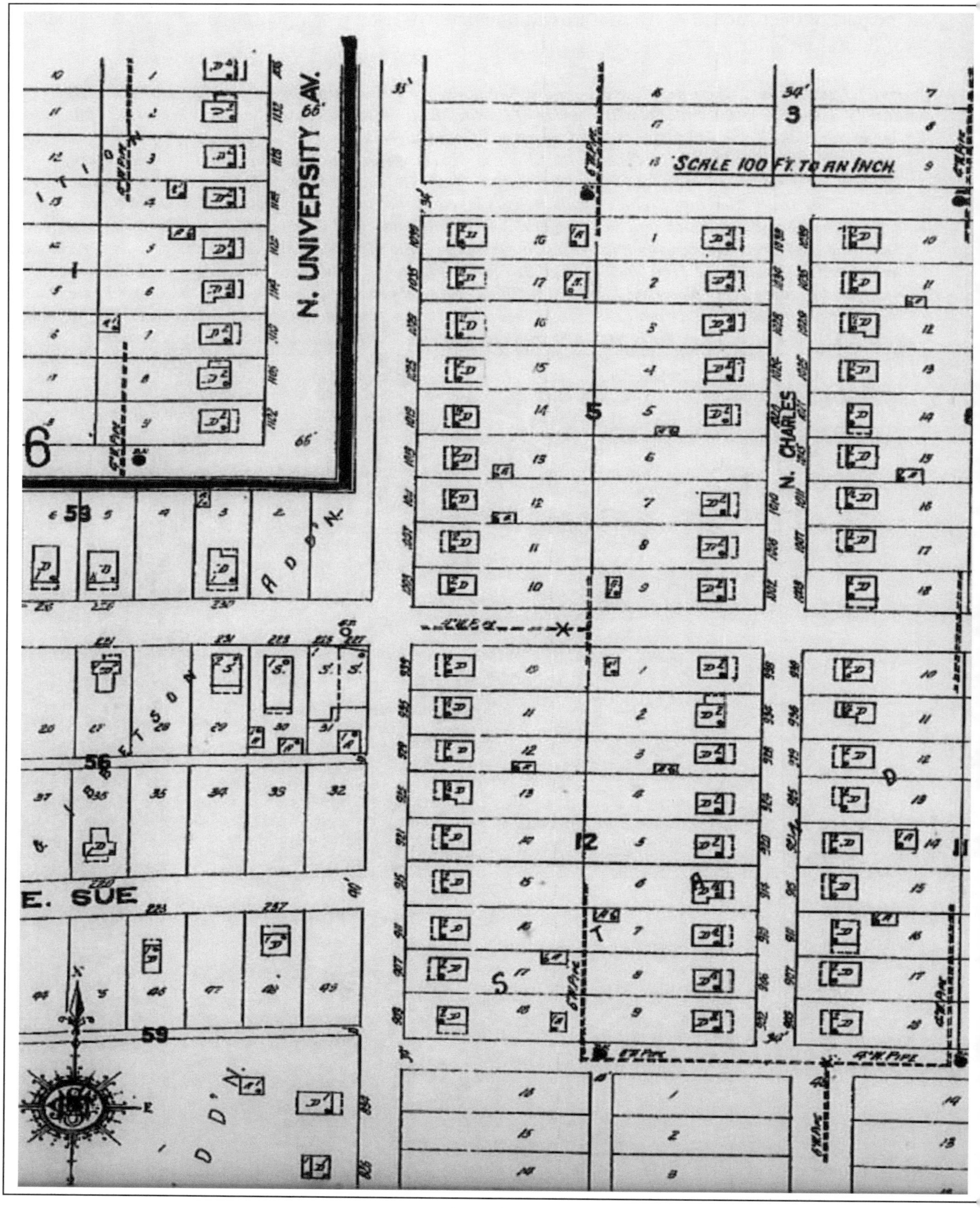

1925 Map of Carlinville Illinois that shows the Standard Additio

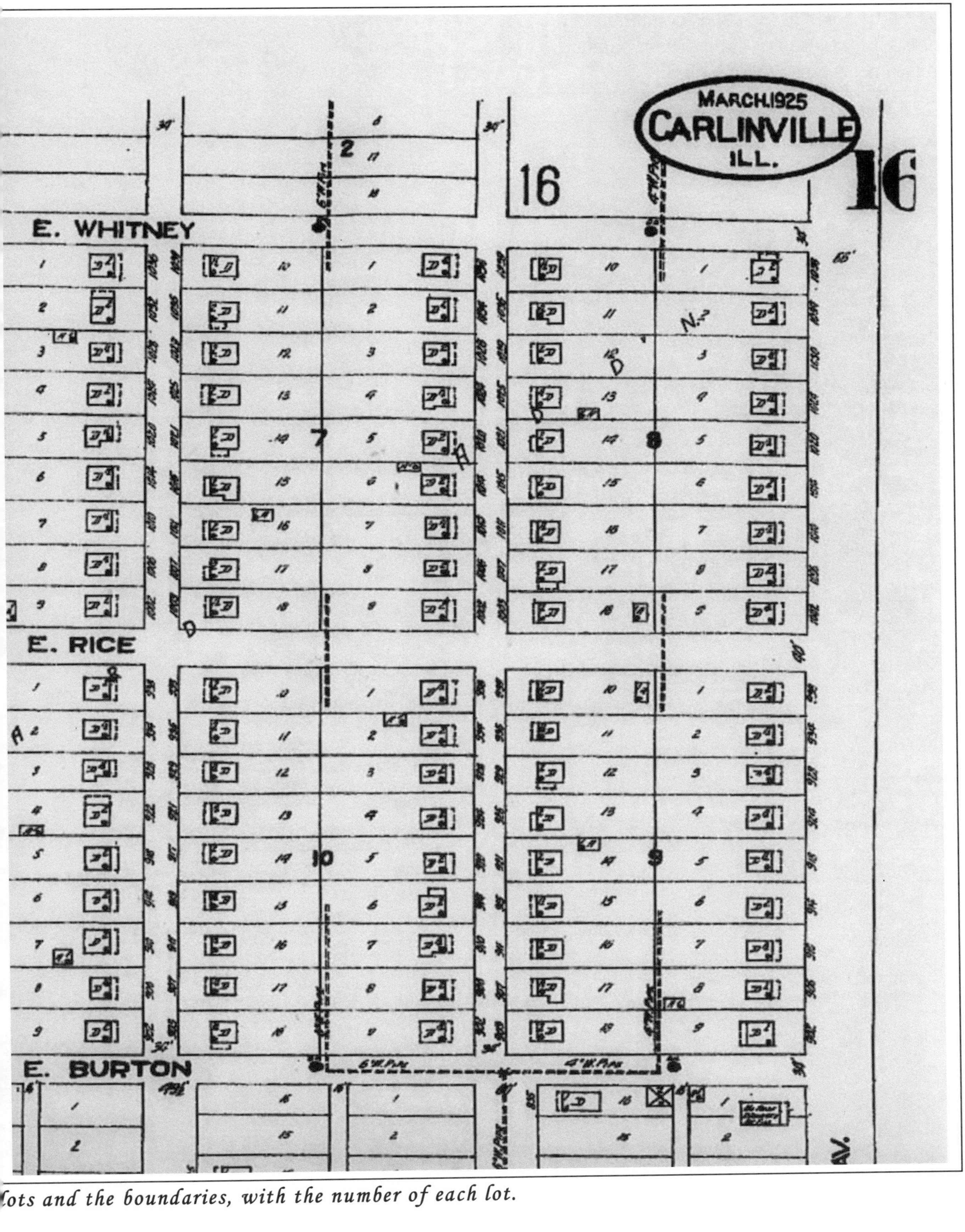

lots and the boundaries, with the number of each lot.

There were three home styles containing 5 rooms - a living room, dining room, kitchen and two bedrooms with a bath.

The Carlin Model

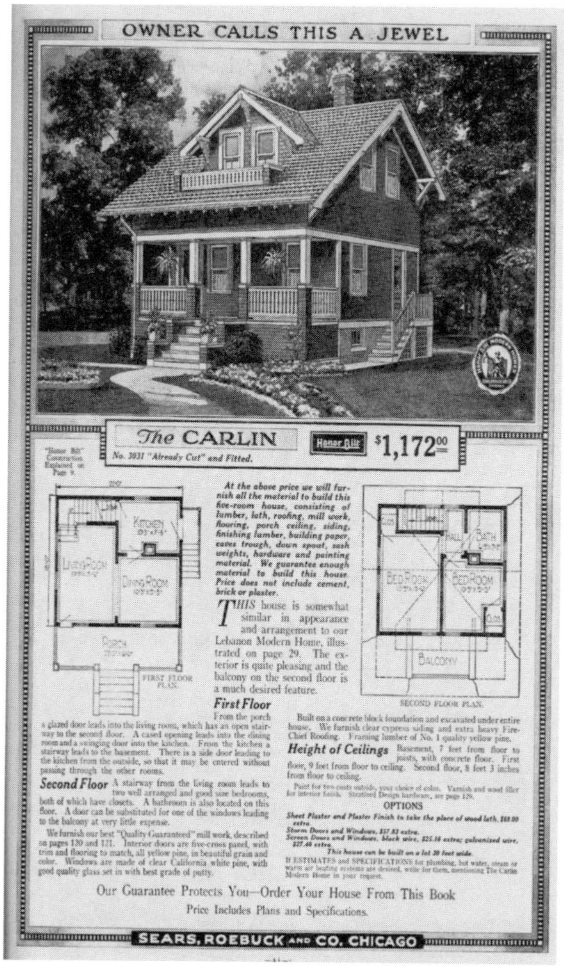

OWNER CALLS THIS A JEWEL

The CARLIN Honor Bilt $1,172.00

No. 3031 "Already Cut" and Fitted.

"Honor Bilt" Construction Explained on Page 9.

At the above price we will furnish all the material to build this five-room house, consisting of lumber, lath, roofing, mill work, flooring, porch ceiling, siding, finishing lumber, building paper, eaves trough, down spout, sash weights, hardware and painting material. We guarantee enough material to build this house. Price does not include cement, brick or plaster.

THIS house is somewhat similar in appearance and arrangement to our Lebanon Modern Home, illustrated on page 29. The exterior is quite pleasing and the balcony on the second floor is a much desired feature.

First Floor From the porch a glazed door leads into the living room, which has an open stairway to the second floor. A cased opening leads into the dining room and a swinging door into the kitchen. From the kitchen a stairway leads to the basement. There is a side door leading to the kitchen from the outside, so that it may be entered without passing through the other rooms.

Second Floor A stairway from the living room leads to two well arranged and good size bedrooms, both of which have closets. A bathroom is also located on this floor. A door can be substituted for one of the windows leading to the balcony at very little expense.

We furnish our best "Quality Guaranteed" mill work, described on pages 120 and 121. Interior doors are five-cross panel, with trim and flooring to match, all yellow pine, in beautiful grain and color. Windows are made of clear California white pine, with good quality glass set in with best grade of putty.

Built on a concrete block foundation and excavated under entire house. We furnish clear cypress siding and extra heavy Fire-Chief Roofing. Framing lumber of No. 1 quality yellow pine.

Height of Ceilings Basement, 7 feet from floor to joists, with concrete floor. First floor, 9 feet from floor to ceiling. Second floor, 8 feet 3 inches from floor to ceiling.

Paint for two coats outside, your choice of color. Varnish and wood filler for interior finish. Stratford Design hardware, see page 129.

OPTIONS

Sheet Plaster and Plaster Finish to take the place of wood lath, $88.00 extra.
Storm Doors and Windows, $57.82 extra.
Screen Doors and Windows, black wire, $25.16 extra; galvanized wire, $27.46 extra.
This house can be built on a lot 30 feet wide.

If ESTIMATES and SPECIFICATIONS for plumbing, hot water, steam or warm air heating systems are desired, write for them, mentioning The Carlin Modern Home in your request.

Our Guarantee Protects You—Order Your House From This Book
Price Includes Plans and Specifications.

SEARS, ROEBUCK AND CO. CHICAGO

The Roseberry Model

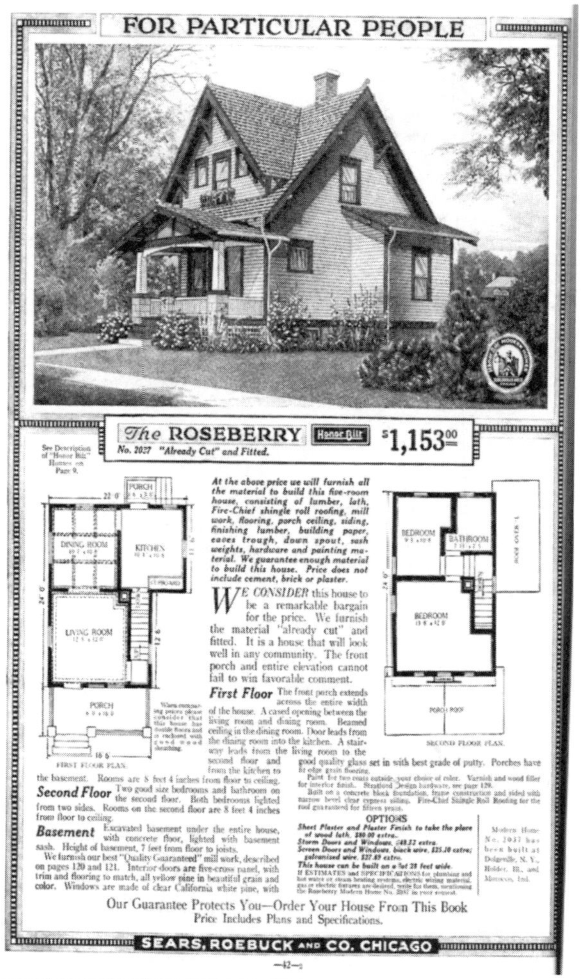

FOR PARTICULAR PEOPLE

The ROSEBERRY Honor Bilt $1,153.00

No. 2027 "Already Cut" and Fitted.

See Description of "Honor Bilt" Homes on Page 9.

At the above price we will furnish all the material to build this five-room house, consisting of lumber, lath, Fire-Chief shingle roll roofing, mill work, flooring, porch ceiling, siding, finishing lumber, building paper, eaves trough, down spout, sash weights, hardware and painting material. We guarantee enough material to build this house. Price does not include cement, brick or plaster.

WE CONSIDER this house to be a remarkable bargain for the price. We furnish the material "already cut" and fitted. It is a house that will look well in any community. The front porch and entire elevation cannot fail to win favorable comment.

When comparing prices please consider that this house has double floors and is enclosed with good wood sheathing.

First Floor The front porch extends across the entire width of the house. A cased opening between the living room and dining room. Beamed ceiling in the dining room. Door leads from the dining room into the kitchen. A stairway leads from the living room to the second floor and from the kitchen to the basement. Rooms are 8 feet 4 inches from floor to ceiling.

Second Floor Two good size bedrooms and bathroom on the second floor. Both bedrooms lighted from two sides. Rooms on the second floor are 8 feet 4 inches from floor to ceiling.

Basement Excavated basement under the entire house, with concrete floor, lighted with basement sash. Height of basement, 7 feet from floor to joists.

We furnish our best "Quality Guaranteed" mill work, described on pages 120 and 121. Interior doors are five-cross panel, with trim and flooring to match, all yellow pine in beautiful grain and color. Windows are made of clear California white pine, with good quality glass set in with best grade of putty. Porches have fir edge grain flooring.

Paint for two coats outside, your choice of color. Varnish and wood filler for interior finish. Stratford Design hardware, see page 129.

Built on a concrete block foundation, frame construction and sided with narrow bevel clear cypress siding. Fire-Chief Shingle Roll Roofing for the roof guaranteed for fifteen years.

OPTIONS

Sheet Plaster and Plaster Finish to take the place of wood lath, $80.00 extra.
Storm Doors and Windows, $48.32 extra.
Screen Doors and Windows, black wire, $25.10 extra; galvanized wire, $27.49 extra.
This house can be built on a lot 28 feet wide.

If ESTIMATES and SPECIFICATIONS for plumbing and hot water or steam heating systems, electric wiring material, gas or electric fixtures are desired, write for them, mentioning the Roseberry Modern Home No. 2027 in your request.

Modern Home No. 2037 has been built at Dolgeville, N. Y., Holder, Ill., and Monroe, Ind.

Our Guarantee Protects You—Order Your House From This Book
Price Includes Plans and Specifications.

SEARS, ROEBUCK AND CO. CHICAGO

—42—

The Lebanon Model

25 SOLD IN ONE MONTH

The LEBANON Honor Bilt $1,092.00

No. 3029 "Already Cut" and Fitted.

At the above price we will furnish all the material to build this five-room house, consisting of lumber, lath, roofing, mill work, flooring, porch ceiling, siding, finishing lumber, building paper, eaves trough, down spout, sash weights, hardware and painting material. We guarantee enough material to build this house. Price does not include cement, brick or plaster.

"Honor Bilt" Construction Explained on Page 9.

FIRST FLOOR PLAN. SECOND FLOOR PLAN.

THIS graceful and nicely arranged house is becoming quite popular with industrial concerns. It solves the house problem where it is desired to retain skilled labor.

First Floor From the porch a glazed door leads into the living room, which has an open stairway to the second floor. A cased opening leads into the dining room and a swinging door into the kitchen, furnished with a convenient pantry case. From the kitchen a stairway leads to the basement. There is a side door leading to the kitchen from the outside, so that it may be entered without passing through the other rooms.

Second Floor A stairway from the living room leads to two well arranged and good size bedrooms, both of which have closets. A bathroom is also located on this floor.

We furnish our best "Quality Guaranteed" mill work, described on pages 120 and 121. Interior doors are five-cross panel, with trim and flooring to match, all yellow pine, in beautiful grain and color. Windows are made of clear California white pine, with good quality glass set in with best grade of putty.

Built on a concrete block foundation and excavated under entire house. We furnish clear cypress siding and extra heavy Fire-Chief Roofing; framing lumber of No. 1 quality yellow pine.

Height of Ceilings Basement, 7 feet from floor to joists, with concrete floor. First floor, 9 feet from floor to ceiling. Second floor, 8 feet 3 inches from floor to ceiling.

Paint for two coats outside, your choice of color. Varnish and wood filler for interior finish. Stratford Design hardware, see page 129.

OPTIONS

Sheet Plaster and Plaster Finish to take the place of wood lath, $67.00 extra.
Storm Doors and Windows, $52.22 extra.
Screen Doors and Windows, black wire, $24.34 extra; galvanized wire, $26.53 extra.
This house can be built on a lot 30 feet wide.

IF ESTIMATES and SPECIFICATIONS for plumbing, hot water, steam or warm air heating systems, electric wiring material, gas or electric fixtures are desired, write for them, mentioning the Lebanon Modern Home No. 3029 in your request.

Our Guarantee Protects You—Order Your House From This Book
Price Includes Plans and Specifications.

SEARS, ROEBUCK AND CO. CHICAGO

—29—

There were 5 different home styles containing 6 rooms - a living room, dining room, kitchen and three bedrooms with a bath.

The Madelia Model

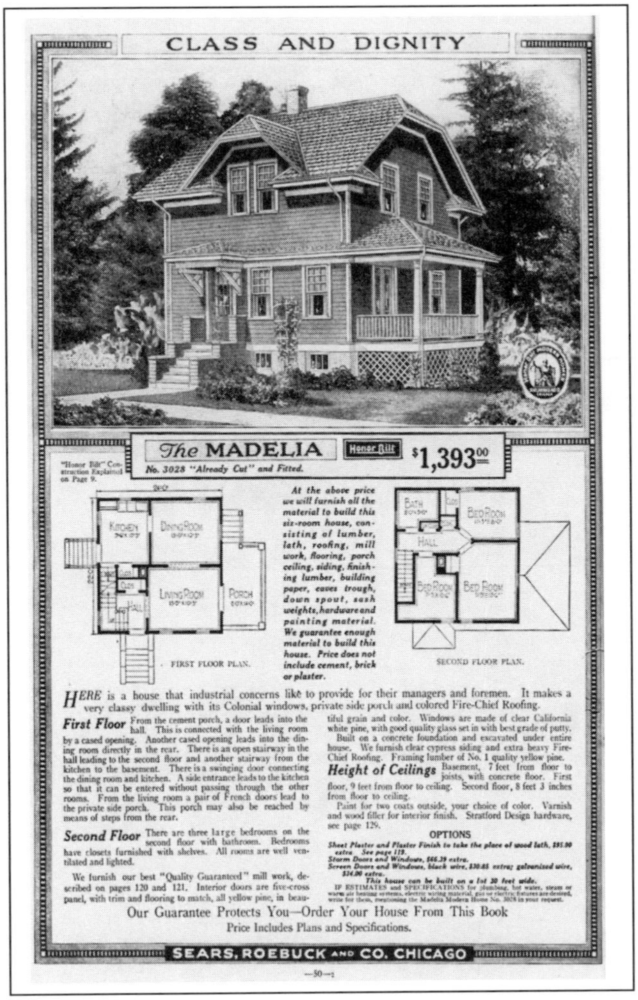

CLASS AND DIGNITY

The MADELIA Honor Bilt $1,393.00

No. 3028 "Already Cut" and Fitted.

"Honor Bilt" Construction Explained on Page 9.

At the above price we will furnish all the material to build this six-room house, consisting of lumber, lath, roofing, mill work, flooring, porch ceiling, siding, finishing lumber, building paper, eaves trough, down spout, sash weights, hardware and painting material. We guarantee enough material to build this house. Price does not include cement, brick or plaster.

FIRST FLOOR PLAN. SECOND FLOOR PLAN.

HERE is a house that industrial concerns like to provide for their managers and foremen. It makes a very classy dwelling with its Colonial windows, private side porch and colored Fire-Chief Roofing.

First Floor From the cement porch, a door leads into the hall. This is connected with the living room by a cased opening. Another cased opening leads into the dining room directly in the rear. There is an open stairway in the hall leading to the second floor and another stairway from the kitchen to the basement. There is a swinging door connecting the dining room and kitchen. A side entrance leads to the kitchen so that it can be entered without passing through the other rooms. From the living room a pair of French doors lead to the private side porch. This porch may also be reached by means of steps from the rear.

Second Floor There are three large bedrooms on the second floor with bathroom. Bedrooms have closets furnished with shelves. All rooms are well ventilated and lighted.

We furnish our best "Quality Guaranteed" mill work, described on pages 120 and 121. Interior doors are five-cross panel, with trim and flooring to match, all yellow pine, in beautiful grain and color. Windows are made of clear California white pine, with good quality glass set in with best grade of putty.

Built on a concrete foundation and excavated under entire house. We furnish clear cypress siding and extra heavy Fire-Chief Roofing. Framing lumber of No. 1 quality yellow pine.

Height of Ceilings Basement, 7 feet from floor to joists, with concrete floor. First floor, 9 feet from floor to ceiling. Second floor, 8 feet 3 inches from floor to ceiling.

Paint for two coats outside, your choice of color. Varnish and wood filler for interior finish. Stratford Design hardware, see page 129.

OPTIONS

Sheet Plaster and Plaster Finish to take the place of wood lath, $95.00 extra. See page 119.
Storm Doors and Windows, $66.39 extra.
Screen Doors and Windows, black wire, $30.85 extra; galvanized wire, $34.00 extra.
This house can be built on a lot 30 feet wide.

IF ESTIMATES and SPECIFICATIONS for plumbing, hot water, steam or warm air heating systems, electric wiring material, gas or electric fixtures are desired, write for them, mentioning the Madelia Modern Home No. 3028 in your request.

Our Guarantee Protects You—Order Your House From This Book
Price Includes Plans and Specifications.

SEARS, ROEBUCK AND CO. CHICAGO

—30—

☙ The Gladstone Model ❧

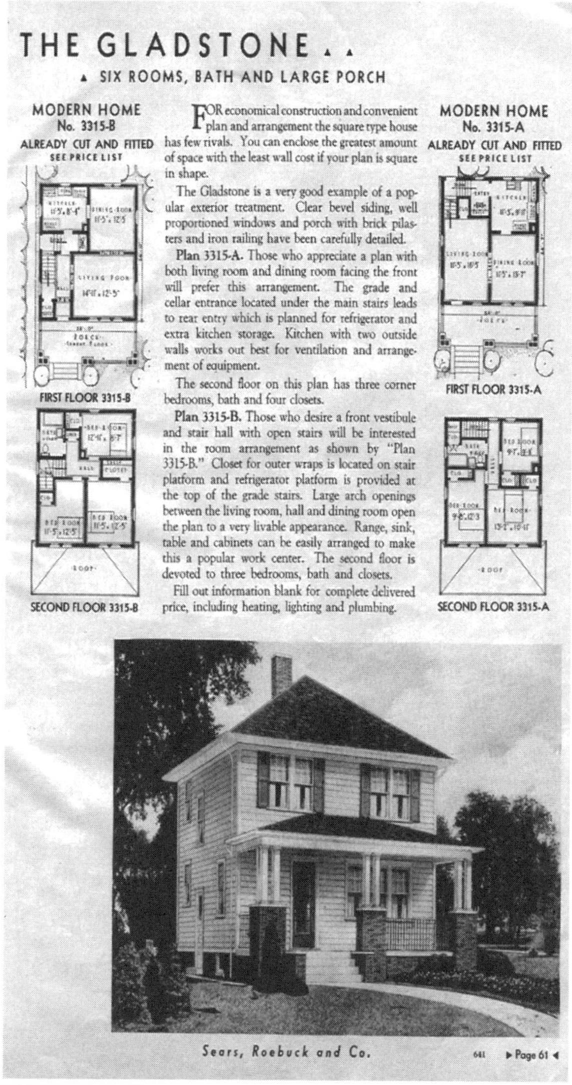

THE GLADSTONE . .

. SIX ROOMS, BATH AND LARGE PORCH

MODERN HOME
No. 3315-B
ALREADY CUT AND FITTED
SEE PRICE LIST

FOR economical construction and convenient plan and arrangement the square type house has few rivals. You can enclose the greatest amount of space with the least wall cost if your plan is square in shape.

The Gladstone is a very good example of a popular exterior treatment. Clear bevel siding, well proportioned windows and porch with brick pilasters and iron railing have been carefully detailed.

Plan 3315-A. Those who appreciate a plan with both living room and dining room facing the front will prefer this arrangement. The grade and cellar entrance located under the main stairs leads to rear entry which is planned for refrigerator and extra kitchen storage. Kitchen with two outside walls works out best for ventilation and arrangement of equipment.

The second floor on this plan has three corner bedrooms, bath and four closets.

Plan 3315-B. Those who desire a front vestibule and stair hall with open stairs will be interested in the room arrangement as shown by "Plan 3315-B." Closet for outer wraps is located on stair platform and refrigerator platform is provided at the top of the grade stairs. Large arch openings between the living room, hall and dining room open the plan to a very livable appearance. Range, sink, table and cabinets can be easily arranged to make this a popular work center. The second floor is devoted to three bedrooms, bath and closets.

Fill out information blank for complete delivered price, including heating, lighting and plumbing.

MODERN HOME
No. 3315-A
ALREADY CUT AND FITTED
SEE PRICE LIST

FIRST FLOOR 3315-B

FIRST FLOOR 3315-A

SECOND FLOOR 3315-B

SECOND FLOOR 3315-A

Sears, Roebuck and Co. 641 ▸ Page 61 ◂

❧

☙ Inside The Gladstone Model ❧

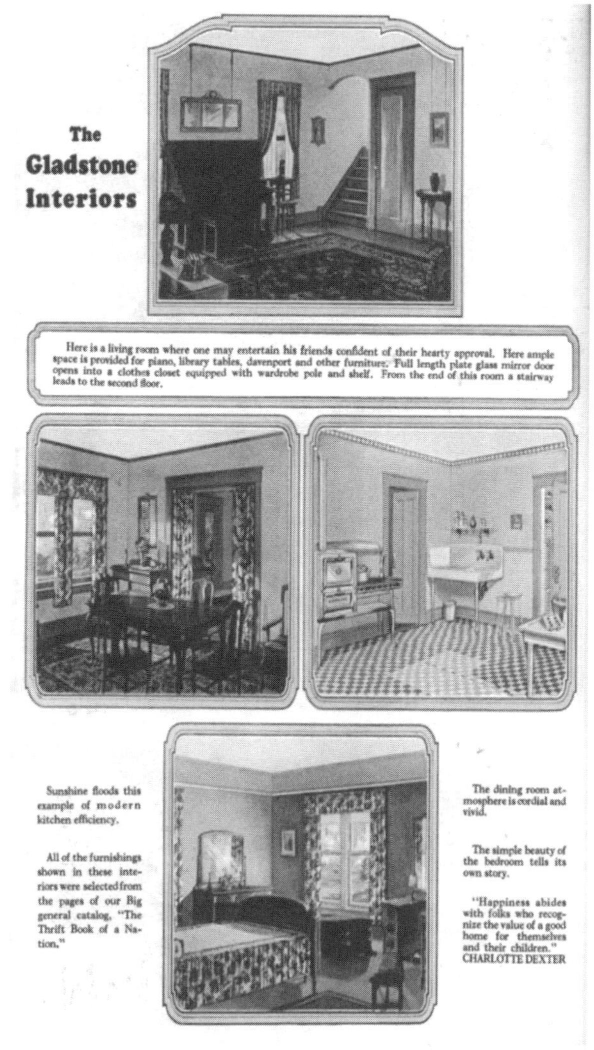

The Gladstone Interiors

Here is a living room where one may entertain his friends confident of their hearty approval. Here ample space is provided for piano, library tables, davenport and other furniture. Full length plate glass mirror door opens into a clothes closet equipped with wardrobe pole and shelf. From the end of this room a stairway leads to the second floor.

Sunshine floods this example of modern kitchen efficiency.

All of the furnishings shown in these interiors were selected from the pages of our Big general catalog, "The Thrift Book of a Nation."

The dining room atmosphere is cordial and vivid.

The simple beauty of the bedroom tells its own story.

"Happiness abides with folks who recognize the value of a good home for themselves and their children."
CHARLOTTE DEXTER

❧

The Langston Model

STURDY AND ATTRACTIVE

The LANGSTON Honor Bilt $1,494.00

No. 2000 "Already Cut" and Fitted.

See Description of "Honor Bilt" Houses on Page 9.

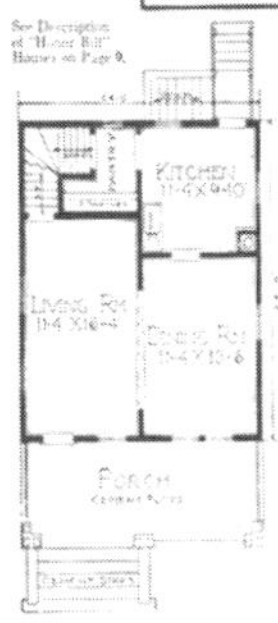

FIRST FLOOR PLAN

At the above price we will furnish all the material to build this six-room house, consisting of lumber, lath, shingles, mill work, flooring, porch ceiling, siding, finishing lumber, building paper, eaves trough, down spout, sash weights, hardware and painting material. We guarantee enough material to build this house. Price does not include cement, brick or plaster.

A *NEAT* and roomy house at a very low price. Was designed especially with two objects in view, economy of floor space and low cost. Contains six rooms and bathroom. Porch is 24 feet by 8 feet.

First Floor Front door, with bevel plate glass, opens into the living room. A stairway from the living room leads to the second floor. The dining room is connected with the living room by a large cased opening which makes practically one large room of these two rooms. Has kitchen and good size pantry. Entrance to the basement is through this pantry.

Second Floor The stairway from the first floor opens into a hall on the second floor from which any of the three bedrooms or bathroom may be reached. All three bedrooms are well lighted and each has a good size clothes closet.

We furnish our best "Quality Guaranteed" mill work, described on pages 120 and 121. Interior doors are five-cross panel, with trim and flooring to match, all yellow pine, in beautiful grain and color. Windows are made of clear California white pine, with good quality glass set in with best grade of putty. Front porch has concrete floor.

Built on a concrete block foundation and excavated under entire house. We furnish clear cypress siding and best grade of thick cedar shingles; framing lumber of No. 1 quality yellow pine.

Height of Ceilings Basement, 7 feet from floor to joists, with concrete floor. First floor, 9 feet from floor to ceiling. Second floor, 8 feet 3 inches from floor to ceiling.

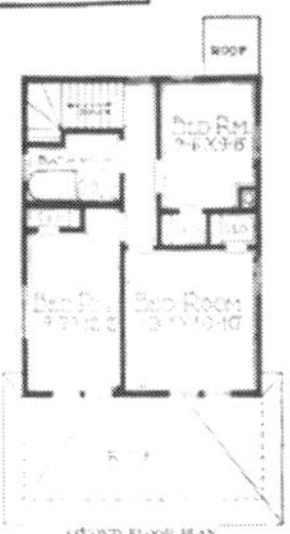

SECOND FLOOR PLAN

Paint for two coats outside, your choice of color. Varnish and wood filler for interior finish. Stratford Design hardware, see page 129.

OPTIONS

***Sheet Plaster and Plaster Finish** to take the place of wood lath, $115.00 extra.*
***Fire-Chief Shingle Roll Roofing**, Red or Sea Green in color, instead of cedar shingles, $34.00 less.*
***Oak Doors, Trim and Floors** in living room and dining room instead of yellow pine, $33.00 extra.*
***Maple Flooring** for kitchen, pantry and bathroom instead of yellow pine, no extra charge.*
***Wood Floor** material for the porch, $40.00 extra.*
***Storm Doors and Windows**, $74.90 extra.*
***Screen Doors and Windows**, black wire, $33.11 extra; galvanized wire, $36.24 extra.*
This house can be built on a lot 30 feet wide.

If ESTIMATES and SPECIFICATIONS for plumbing, hot water, steam or warm air heating systems, electric wiring material, gas or electric fixtures are desired, write for them, mentioning the Langston Modern Home No. 2000 in your request.

This house has been built at Roanoke, Ind., Dayton, Ohio, Detroit, Mich., Fort Wayne, Ind., Kankakee, Ill., Wellsville, Ohio, and other cities.

Our Guarantee Protects You—Order Your House From This Book

Price Includes Plans and Specifications.

SEARS, ROEBUCK AND CO. CHICAGO

Inside The Langston Model

INTERIORS - *The* LANGSTON

The Living Room

THIS photographic reproduction will help you in the interior treatment of your new home. Many other ideas are yours for the asking.

The Dining Room

THIS room looks dignified in its simplicity. It is rich and up to date without being loud.

A GLANCE at the interiors of THE LANGSTON MODERN HOME illustrated above and the exterior on the opposite page should convince you of its value as an investment. It will make you a handsome, durable home for your own use or you can sell it quickly at a big profit if you so desire.

THE LANGSTON has been built at Roanoke, Ind.; Dayton, Ohio; Detroit, Mich.; Fort Wayne, Ind.; Kankakee, Ill.; Wellsville, Ohio, and other cities. Ask us for names of Modern Home owners in your own state. There may be some in your own county or even in your own town. They will put you on the right track in building to advantage at small cost.

SEARS, ROEBUCK AND CO. CHICAGO

—53—

The Gladstone and Langston Models are identical except for the dormer at the top. The Gladstone could be built without the dormer on the roof and most were. The Langston always had to be built with the dormer.

The Langston was model number 2000 in **1918** with the dormer on the roof. The Gladstone did not make its debut in the Sears Honor Bilt Modern Homes catalog until **1923.** So technically, there really aren't any Gladstones in Standard Addition. This means that they probably came out with the plans and first started building the Gladstone, or Langston without a dormer, in **1918** or **1919**, which is when the houses were built in Standard Addition. Sears' policy was to try out new house model kits for a couple of years before they were offered in the catalog to customers. So therefore, the Gladstone's in the Addition were some of the 1st to ever be built.

Therefore, it only makes sense to call houses that have the dormer in the Addition - Langston Models and the houses without the dormer in the Addition - Gladstone Models. This is what Sears officials decided. Maybe we should just call them Langston/Gladstone Models and leave it at that. What do you think?

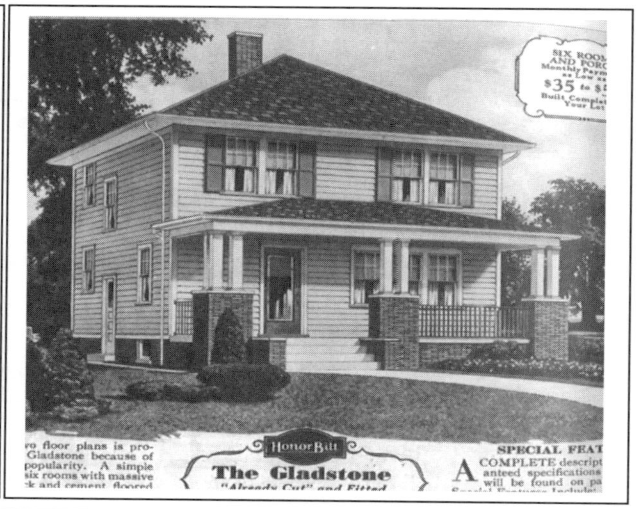

Above a Langston and a Gladstone – you decide.

Left is a view of Standard Addition under construction in ***1918-1919.***

ଔ The Whitehall Model ଓ

NEAT AND ROOMY

The WHITEHALL Honor Bilt $1,465.00

No. 3035 "Already Cut" and Fitted.

See Description of "Honor Bilt" Houses on Page 9.

At the above price we will furnish all the material to build this six-room house, consisting of lumber, lath, shingles, mill work, flooring, porch ceiling, siding, finishing lumber, building paper, eaves trough, down spout, sash weights, hardware and painting material. No extras, as we guarantee enough material to build this house. Price does not include cement, brick or plaster.

A NEAT and roomy house at a very low price. Was designed with two objects in view, economy of floor space and low cost. Contains six good size rooms and bathroom. Front door opens into the living room. A stairway from the living room leads to the second floor. The dining room is connected with the living room by a large cased opening which practically makes one large room of these two rooms. Has kitchen and good sized pantry. Inside entrance to the basement.

First Floor The front door is 1¾ inches thick, glazed with bevel plate glass. Living room, dining room, kitchen and pantry on first floor. **Double floors** in all rooms. Porch floor is clear edge grain fir.

Second Floor The stairway from the first floor opens into a hall on the second floor from which any one of the three bedrooms or bathroom may be reached. Bedrooms are well lighted and each has a good size clothes closet.

Height of Ceilings Basement, 7 feet from floor to joists, with cement floor. First floor, 9 feet from floor to ceiling. Second floor, 8 feet 6 inches from floor to ceiling.

We furnish our best "Quality Guaranteed" mill work, described on pages 120 and 121. Interior doors are five-cross panel, with trim and flooring to match, all yellow pine, in beautiful grain and color. Windows are made of clear California white pine, with good quality glass set in with best grade of putty. Porches have fir edge grain flooring.

Built on a concrete foundation and excavated under entire house. We furnish clear cypress siding over good wood sheathing and best grade thick cedar shingles; framing lumber is clear quality yellow pine. All windows glazed with "A" quality glass.

Paint for two coats outside. Varnish and wood filler for interior finish. Stratford Design hardware, see page 129.

OPTIONS

Sheet Plaster and Plaster Finish to take the place of wood lath, $104.00 extra.

Fire-Chief Shingle Roll Roofing, Red or Sea Green in color, instead of wood shingles, $28.00 less.

Storm Doors and Windows, $77.42 extra.

Screen Doors and Windows, black wire, $35.20 extra; galvanized wire, $38.71 extra.

This house can be built on a lot 28 feet wide.

If ESTIMATES and SPECIFICATIONS for plumbing, hot water, steam or warm air heating systems, electric wiring material, gas or electric fixtures are desired, write for them, mentioning the Whitehall Modern Home No. 3035 in your request.

Built at Gary, Ind., Davenport, Iowa, Laporte, Ind., Moorestown, N. J., Eastwood, N. Y., Hadley, Penn., Cresco, Iowa, and West Point, Va.

Our Guarantee Protects You—Order Your House From This Book
Price Includes Plans and Specifications.

SEARS, ROEBUCK AND CO. CHICAGO

ଓ

ଔ The Whitehall Model Blueprints ଓ

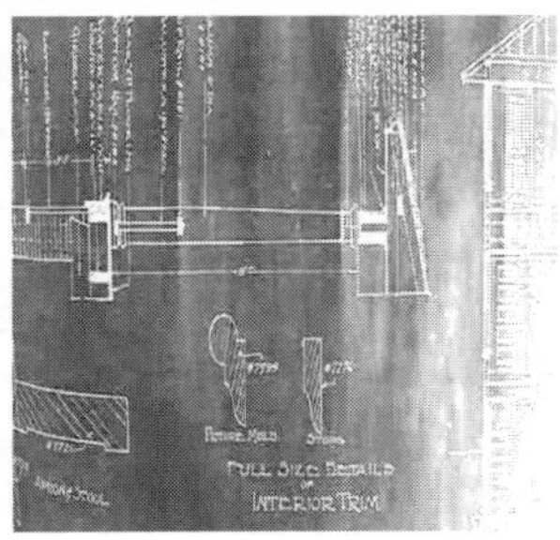

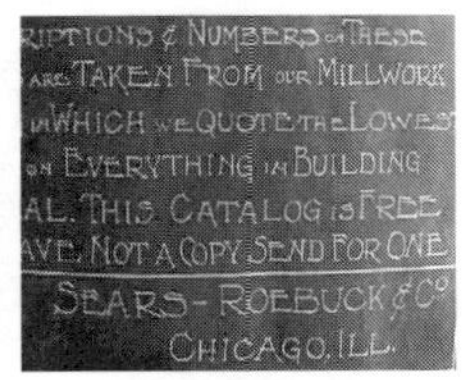

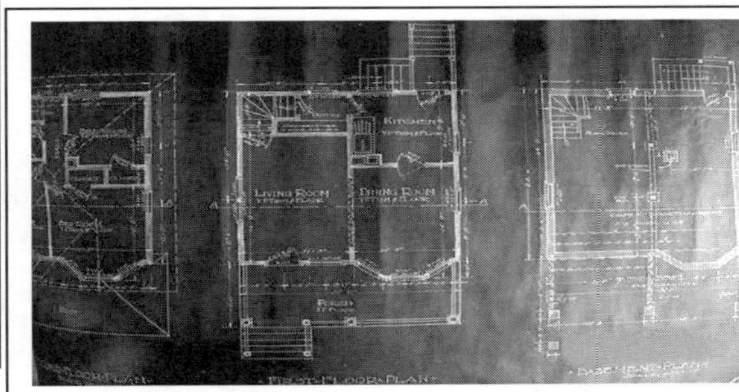

ଓ

The Warrenton Model

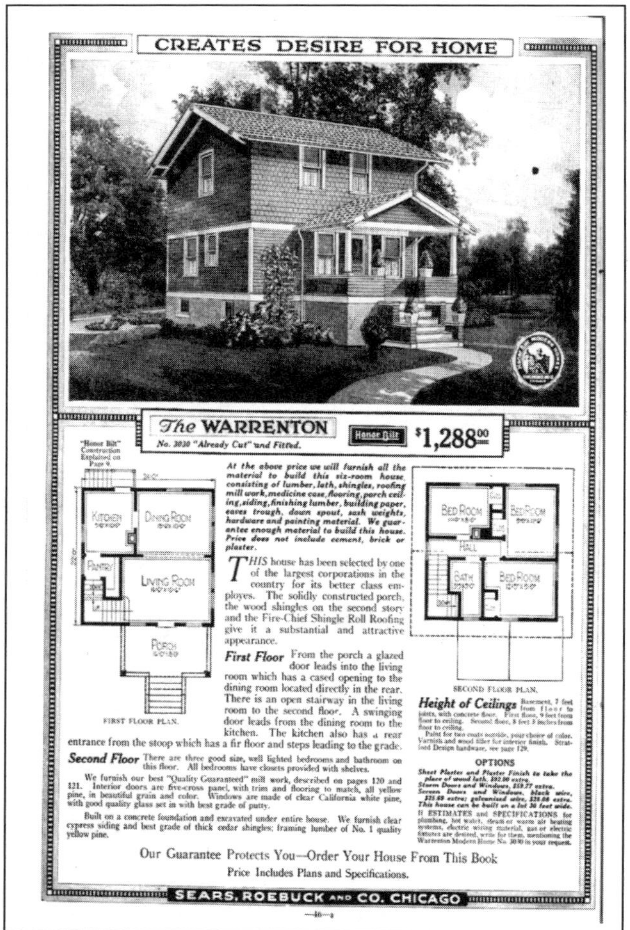

CREATES DESIRE FOR HOME

The WARRENTON

No. 3030 "Already Cut" and Fitted.

Honor Bilt

$1,288.00

"Honor Bilt" Construction Explained on Page 9.

At the above price we will furnish all the material to build this six-room house, consisting of lumber, lath, shingles, roofing, mill work, medicine case, flooring, porch ceiling, siding, finishing lumber, building paper, eaves trough, down spout, sash weights, hardware and painting material. We guarantee enough material to build this house. Price does not include cement, brick or plaster.

THIS house has been selected by one of the largest corporations in the country for its better class employes. The solidly constructed porch, the wood shingles on the second story and the Fire-Chief Shingle Roll Roofing give it a substantial and attractive appearance.

First Floor From the porch a glazed door leads into the living room which has a cased opening to the dining room located directly in the rear. There is an open stairway in the living room to the second floor. A swinging door leads from the dining room to the kitchen. The kitchen also has a rear entrance from the stoop which has a fir floor and steps leading to the grade.

Second Floor There are three good size, well lighted bedrooms and bathroom on this floor. All bedrooms have closets provided with shelves.

We furnish our best "Quality Guaranteed" mill work, described on pages 120 and 121. Interior doors are five-cross panel, with trim and flooring to match, all yellow pine, in beautiful grain and color. Windows are made of clear California white pine, with good quality glass set in with best grade of putty.

Built on a concrete foundation and excavated under entire house. We furnish clear cypress siding and best grade of thick cedar shingles; framing lumber of No. 1 quality yellow pine.

FIRST FLOOR PLAN.

SECOND FLOOR PLAN.

Height of Ceilings Basement, 7 feet from floor to joists, with concrete floor. First floor, 9 feet from floor to ceiling. Second floor, 8 feet 3 inches from floor to ceiling.

Paint for two coats outside, your choice of color. Varnish and wood filler for interior finish. Stratford Design hardware, see page 129.

OPTIONS

Sheet Plaster and Plaster Finish to take the place of wood lath, $92.90 extra.
Storm Doors and Windows, $59.77 extra.
Screen Doors and Windows, black wire, $25.69 extra; galvanized wire, $28.06 extra.
This house can be built on a lot 30 feet wide.

If ESTIMATES and SPECIFICATIONS for plumbing, hot water, steam or warm air heating systems, electric wiring material, gas or electric fixtures are desired, write for them, mentioning the Warrenton Modern Home No. 3030 in your request.

Our Guarantee Protects You—Order Your House From This Book

Price Includes Plans and Specifications.

SEARS, ROEBUCK AND CO. CHICAGO

Two of the Warrenton Models that were built in Standard Addition are turned sideways. One of these is on University Street and one is on High Street. It gives you the illusion that it is a different house style, but it is not; it is definitely the Warrenton.

Above is a Warrenton, on Charles Street – all but two of the Warrenton's in Standard Addition are situated like this one. Left is a "turned Sideways" Warrenton Model,. This home is located on High Street.

The sideways Warrenton's look a lot like the Roanoke Model, but trust me, they are not. They are definitely Warrenton's. The floor plans are different as well as details on the outside of the house, which are proof positive.

The Roanoke Model

☙ The Unknown Model ❧

Then, there is the "Unknown Model." This model totally baffles me. Its exterior is almost identical to the "Roseberry Model" except that it has a square shaped frame while the "Roseberry Model" has an "L" shaped frame where the kitchen is.

The Unknown Model.

The eaves are bracketed like the "Roseberry Model," but it has three bedrooms upstairs where the "Roseberry Model" has only two. There are 14 of these homes in Standard Addition; and I cannot find the style of home that they are. They never appeared in any Sears, Roebuck & Co. Catalogs. I have been trying for 20-years to find out this house model, but to no avail. The only thing that it can be is a cross between two different house styles. The downstairs being the "Warrenton Model" and the upstairs being a modification of the "Cranmore Model."

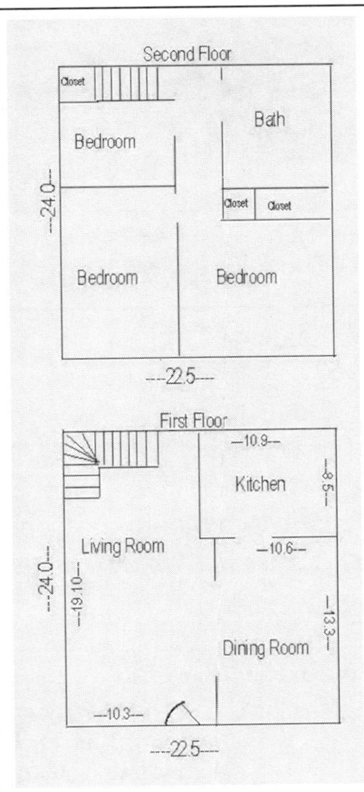

Unknown Model photos of two of the upstairs bedrooms and the downstairs dining room – the mystery of this house style may never be solved.

Debra Spangler Whitlock owns the Unknown Model featured on these pages. Her grandparents, William and Lela Justice, purchased it from M.D. Kessinger in **1947.**

William and Lela Justice – parents of Joann Justice Spangler.

The house has remained in the family ever since. M.D. Kessinger originally purchased the house from Standard Oil Company for $500.00 when they sold off the houses in **1935**.

William and Lela Justice sold the home to Lee and Joann Spangler (Debra's parents)on **July 20, 1967.**

After her father's death, Debra purchased the house on **July 12, 2002**.

Debra was 6 months old when her parents moved into the house. Now age 51, Debra has lived practically her whole life in the home.

A Special Thank You to Debra Whitlock and son Nate, who so graciously allowed me to photograph every corner of their home. The Whitlock's, with their dogs Mac and Rascal, live in the house along with occasional visits and sleepovers by Debra's other son Matt and the grandkids Austin, Zaiden & Ayden.

Right is a view from the front door inside the Unknown Model.

*Below are Lee and Joann Spangler behind their home in the 900 block of Standard Addition in **1950**.*

***1918 –1919** Construction of Sears, Roebuck & Co. Homes in Carlinville, Illinois.*

When all was said and done, Standard had placed a 192-house order with Sears. 156 of these were built in Standard Addition, 12 were built at Schoper and the rest were erected in Wood River, IL across from the Standard Oil Refinery.

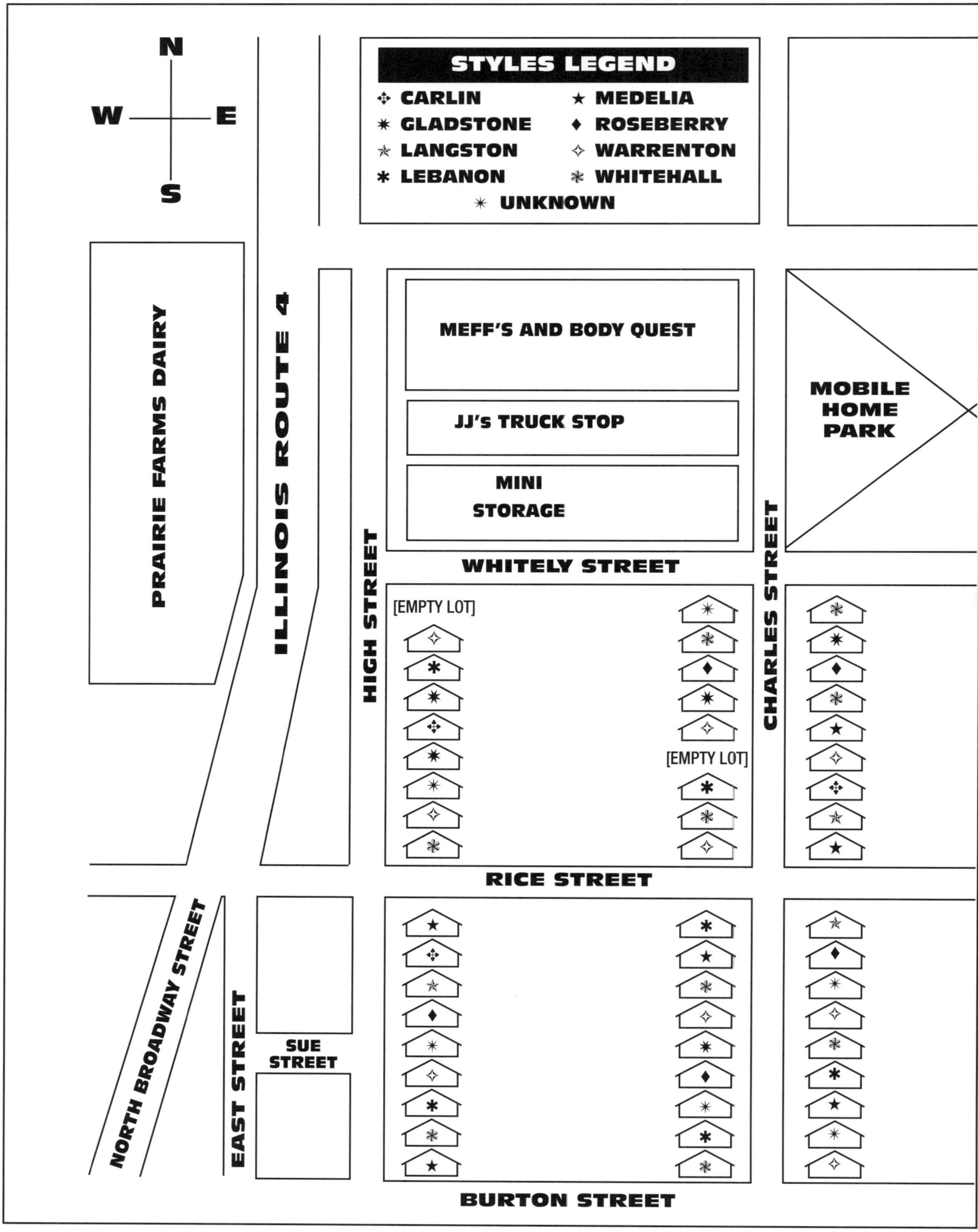

Historic Standard Addition Map – this map shows

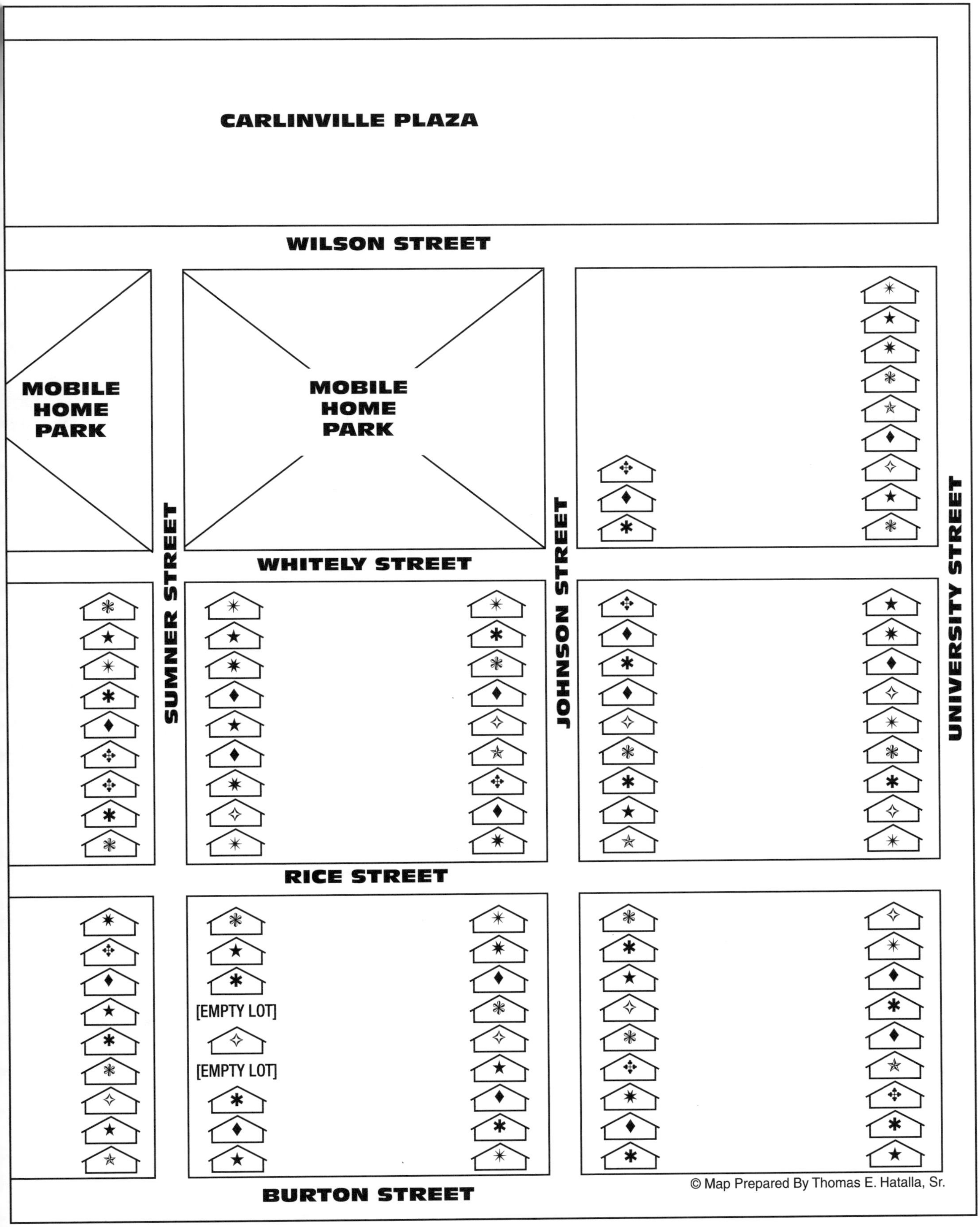

where each style of home is located on each block.

Right are photos of Standard Addition Construction in ***1918 –1919.***

The construction in Standard Addition started on University Street and ended on High Street.

The building of the homes in Standard Addition took 8 ½ months, which is remarkable considering there were few power tools and most of the work was done by hand; but bear in mind that each house was already pre-cut, that definitely made construction faster and easier. Standard had thought that the whole project would be done in about 6 months, but they weren't counting on a wheat harvest.

Right are some of the newly completed homes in Standard Addition in ***1919.***

The Carlin Model was said to have been named after Carlinville by Sears, in honor of the town where so many of their homes were being built. This is most likely true, as the Carlin Model did not debut in the Sears Modern Homes catalog until **1919**, which meant that the plans for it were drawn out in **1918** when Standard Addition was being planned and built.

☙ Would Addition Construction Be Delayed Again? ❧

The building of mining homes was allowed to proceed even though most building projects in the United States were brought to a halt by Chief of the War Industries Board at Washington D.C.

❧ On **October 16, 1918,** C. T. Woodward, county chairman of National Defense received instructions for what projects were allowed to be finished or built. Cottages to house coal miners and their families definitely received the okay to go ahead as the nation needed help increasing the output of coal for heating homes and for the World War I effort.

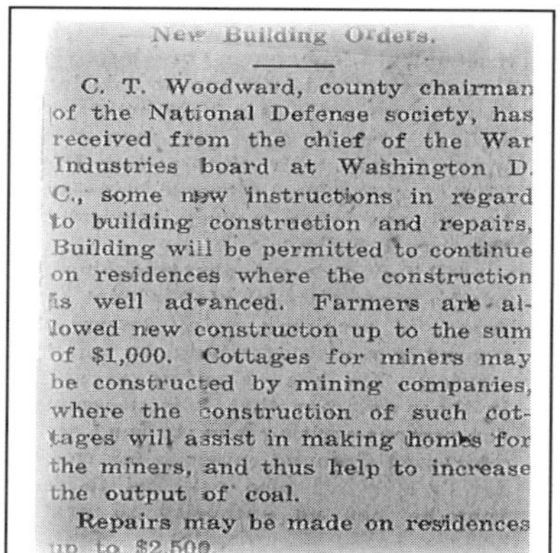

New Building Orders.

C. T. Woodward, county chairman of the National Defense society, has received from the chief of the War Industries board at Washington D. C., some new instructions in regard to building construction and repairs. Building will be permitted to continue on residences where the construction is well advanced. Farmers are allowed new constructon up to the sum of $1,000. Cottages for miners may be constructed by mining companies, where the construction of such cottages will assist in making homes for the miners, and thus help to increase the output of coal.

Repairs may be made on residences up to $2,500.

Addition To Get Sewer System

On **November 4, 1918,** Carlinville City Council approved the request by Standard to put down sewers in their new Addition. The company would bear all of the expenses to do this. The council also voted to extend the water mains on all of the streets leading into the Addition.

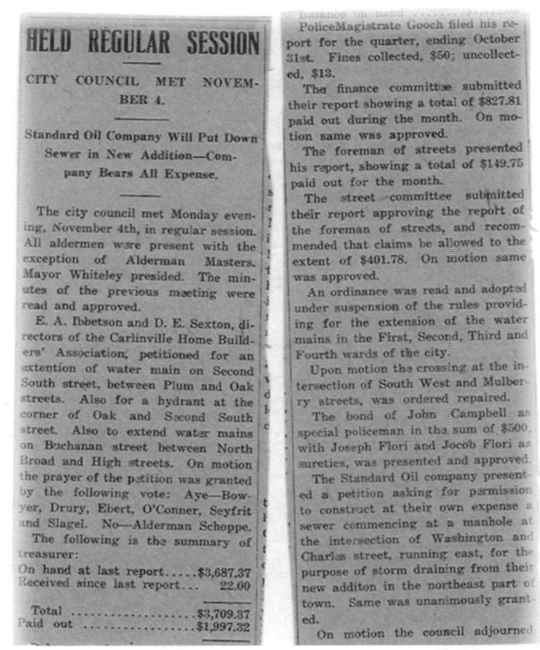

HELD REGULAR SESSION

CITY COUNCIL MET NOVEMBER 4.

Standard Oil Company Will Put Down Sewer in New Addition—Company Bears All Expense.

The city council met Monday evening, November 4th, in regular session. All aldermen were present with the exception of Alderman Masters. Mayor Whiteley presided. The minutes of the previous meeting were read and approved.

E. A. Ibbetson and D. E. Sexton, directors of the Carlinville Home Builders' Association, petitioned for an extention of water main on Second South street, between Plum and Oak streets. Also for a hydrant at the corner of Oak and Second South street. Also to extend water mains on Buchanan street between North Broad and High streets. On motion the prayer of the petition was granted by the following vote: Aye—Bowyer, Drury, Ebert, O'Conner, Seyfrit and Slagel. No—Alderman Schoppe.

The following is the summary of treasurer:

On hand at last report.....$3,687.37
Received since last report... 22.00

Total $3,709.37
Paid out $1,997.32

PoliceMagistrate Gooch filed his report for the quarter, ending October 31st. Fines collected, $50; uncollected, $13.

The finance committee submitted their report showing a total of $827.81 paid out during the month. On motion same was approved.

The foreman of streets presented his report, showing a total of $149.75 paid out for the month.

The street committee submitted their report approving the report of the foreman of streets, and recommended that claims be allowed to the extent of $401.78. On motion same was approved.

An ordinance was read and adopted under suspension of the rules providing for the extension of the water mains in the First, Second, Third and Fourth wards of the city.

Upon motion the crossing at the intersection of South West and Mulberry streets, was ordered repaired.

The bond of John Campbell as special policeman in the sum of $500, with Joseph Flori and Jocob Flori as sureties, was presented and approved.

The Standard Oil company presented a petition asking for permission to construct at their own expense a sewer commencing at a manhole at the intersection of Washington and Charles street, running east, for the purpose of storm draining from their new additon in the northeast part of town. Same was unanimously granted.

On motion the council adjourned

Note: Even though Standard was going to foot the bill for the sewers in the Addition, the water mains and connection had to be extended a quarter of a mile to reach the Addition. It was Carlinville and ultimately, the taxpayers, who bore that expense.

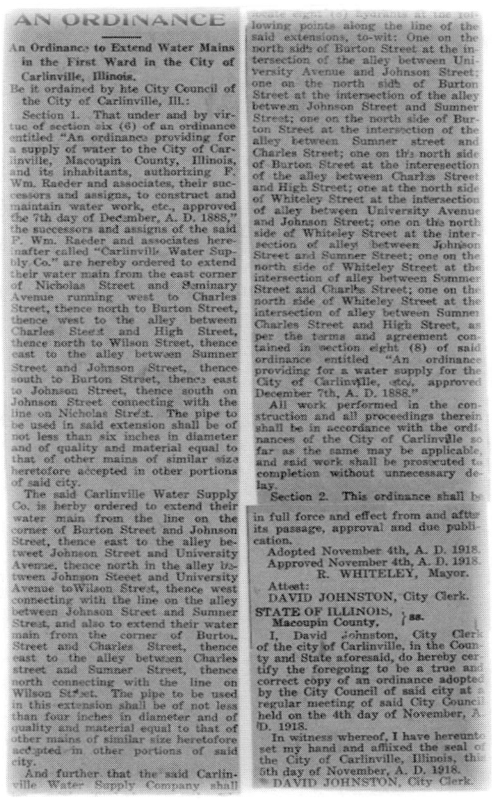

AN ORDINANCE

An Ordinance to Extend Water Mains in the First Ward in the City of Carlinville, Illinois.

Be it ordained by hte City Council of the City of Carlinville, Ill.:

Section 1. That under and by virtue of section six (6) of an ordinance entitled "An ordinance providing for a supply of water to the City of Carlinville, Macoupin County, Illinois, and its inhabitants, authorizing F. Wm. Raeder and associates, their successors and assigns, to construct and maintain water work, etc., approved the 7th day of December, A. D. 1888," the successors and assigns of the said F. Wm. Raeder and associates hereinafter called "Carlinville Water Supbly Co." are hereby ordered to extend their water main from the east corner of Nicholas Street and Seminary Avenue running west to Charles Street, thence north to Burton Street, thence west to the alley between Charles Steet and High Street, thence north to Wilson Street, thence east to the alley between Sumner Street and Johnson Street, thence south to Burton Street, thence east to Johnson Street, thence south on Johnson Street connecting with the line on Nicholas Street. The pipe to be used in said extension shall be of not less than six inches in diameter and of quality and material equal to that of other mains of similar size heretofore accepted in other portions of said city.

The said Carlinville Water Supply Co. is herby ordered to extend their water main from the line on the corner of Burton Street and Johnson Street, thence east to the alley betweet Johnson Street and University Avenue, thence north in the alley between Johnson Steeet and University Avenue toWilson Street, thence west connecting with the line on the alley between Johnson Street and Sumner Street, and also to extend their water main from the corner of Burton Street and Charles Street, thence east to the alley between Charles street and Sumner Street, thence north connecting with the line on Wilson Street. The pipe to be used in this extension shall be of not less than four inches in diameter and of quality and material equal to that of other mains of similar size heretofore accepted in other portions of said city.

And further that the said Carlinville Water Supply Company shall [illegible] eight (8) hydrants at the following points along the line of the said extensions, to-wit: One on the north side of Burton Street at the intersection of the alley between University Avenue and Johnson Street; one on the north side of Burton Street at the intersection of the alley between Johnson Street and Sumner Street; one on the north side of Burton Street at the intersection of the alley between Sumner street and Charles Street; one on the north side of Burton Street at the interesection of the alley between Charles Street and High Street; one at the north side of Whiteley Street at the intersection of alley between University Avenue and Johnson Street; one on the north side of Whiteley Street at the intersection of alley between Johnson Street and Sumner Street; one on the north side of Whiteley Street at the intersection of alley between Sumner Street and Charles Street; one on the north side of Whiteley Street at the intersection of alley between Sumner Charles Street and High Street, as per the terms and agreement contained in section eight (8) of said ordinance entitled "An ordinance providing for a water supply for the City of Carlinville, etc., approved December 7th, A. D. 1888."

All work performed in the construction and all proceedings therein shall be in accordance with the ordinances of the City of Carlinville so far as the same may be applicable, and said work shall be prosecuted to completion without unnecessary delay.

Section 2. This ordinance shall be in full force and effect from and after its passage, approval and due publication.

Adopted November 4th, A. D. 1918.
Approved November 4th, A. D. 1918.
R. WHITELEY, Mayor.

Attest:
DAVID JOHNSTON, City Clerk.

STATE OF ILLINOIS, }
Macoupin County, } ss.

I, David Johnston, City Clerk of the city of Carlinville, in the County and State aforesaid, do hereby certify the foregoing to be a true and correct copy of an ordinance adopted by the City Council of said city at a regular meeting of said City Council held on the 4th day of November, A. D. 1918.

In witness whereof, I have hereunto set my hand and affixed the seal of the City of Carlinville, Illinois, this 5th day of November, A. D. 1918.
DAVID JOHNSTON, City Clerk.

An ordinance was put into effect **November 16, 1918,** to extend the water mains to Standard Addition so that water would be available to every home. This ordinance also allowed for fire hydrants on all corners in the Addition. Everything was coming into place and it would be but a few months before the homes were completed.

Carlinville Schools Crowded

An article in the **Carlinville Democrat** on **February 26, 1919,** said that Carlinville schools were "*crowded.*" "*Carlinville does not want to have children running in the streets because there is no room in the schools for them. This will be the condition next year unless everything is done to provide room. It will be impossible to provide room fast enough to prevent a crowded condition, but it will be possible to provide room so that all children are able to go to school.*"

Houses For Sale

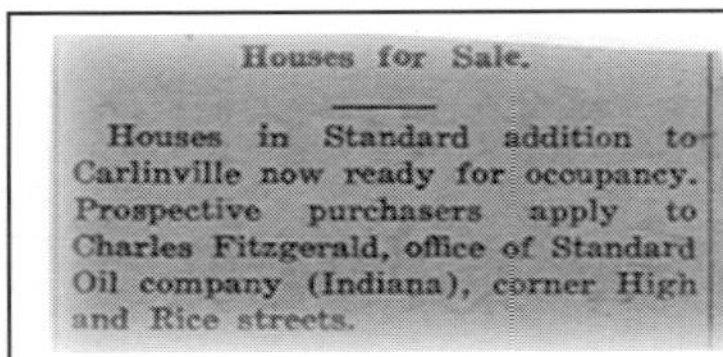

Houses for Sale.

Houses in Standard addition to Carlinville now ready for occupancy. Prospective purchasers apply to Charles Fitzgerald, office of Standard Oil company (Indiana), corner High and Rice streets.

April 23, 1919, the big day had finally arrived, the houses were ready for occupancy and the local paper ran ads proclaiming the news. For several weeks the local paper would run the *"Houses For Sale"* ad. Carlinville's Historic Standard Addition was now completed with all of the modern conveniences. It was time for the people to come, and they did come. In five short weeks every home in Standard Addition was occupied.

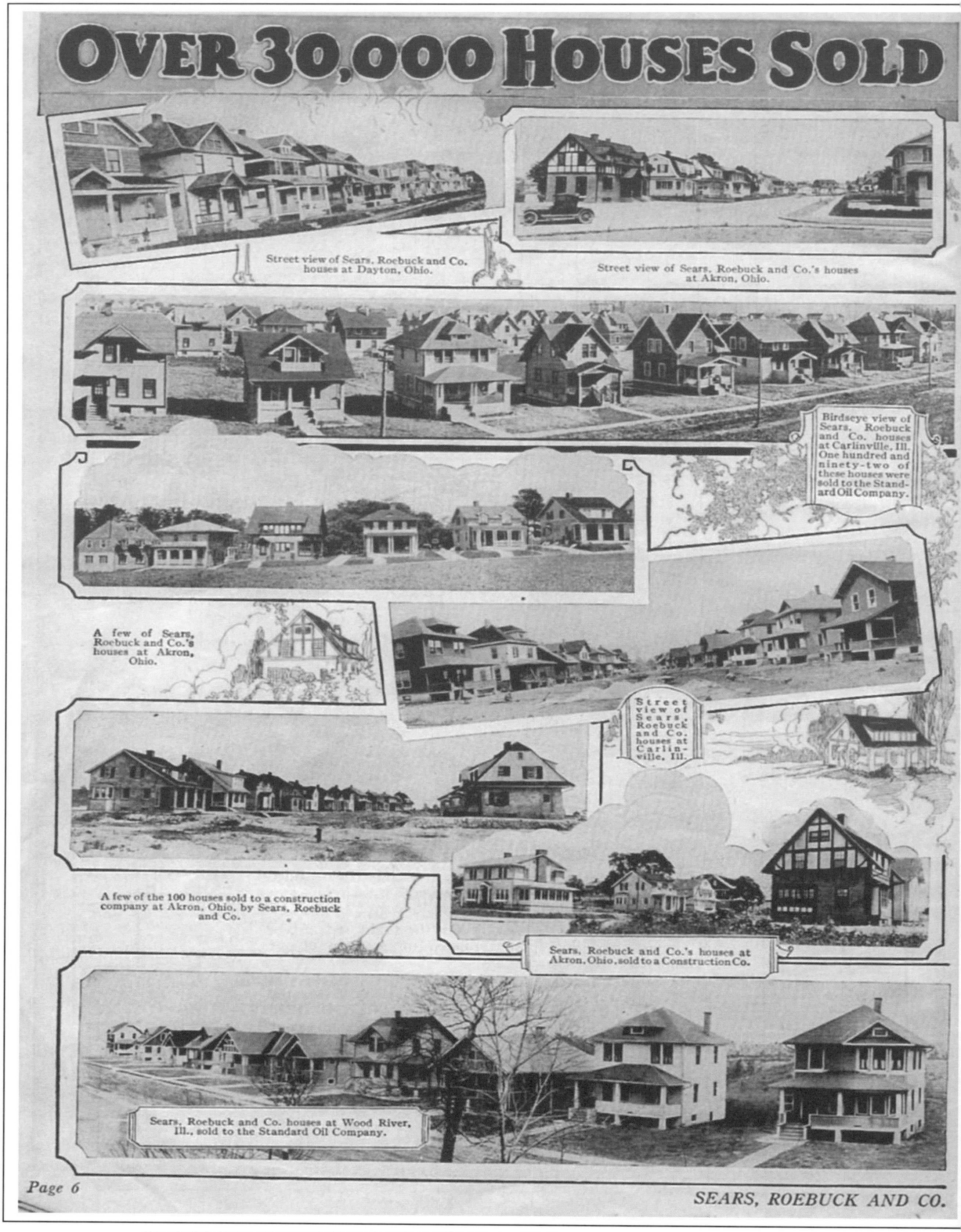

*Double page pictures from the **1922** Sears, Roebuck & Co. Honor Bilt Modern Homes Catalog*

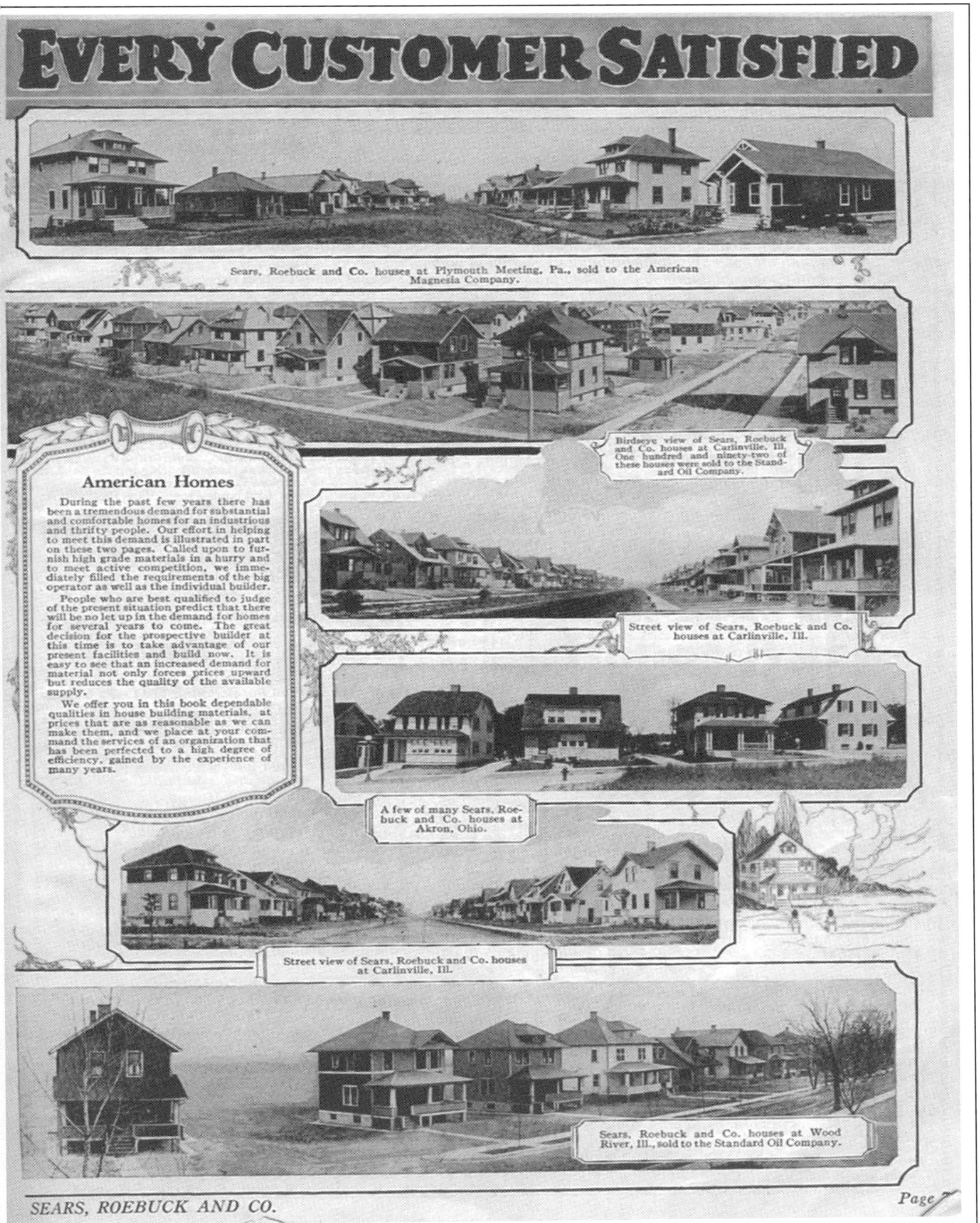

EVERY CUSTOMER SATISFIED

Sears, Roebuck and Co. houses at Plymouth Meeting, Pa., sold to the American Magnesia Company.

Birdseye view of Sears, Roebuck and Co. houses at Carlinville, Ill. One hundred and ninety-two of these houses were sold to the Standard Oil Company.

American Homes

During the past few years there has been a tremendous demand for substantial and comfortable homes for an industrious and thrifty people. Our effort in helping to meet this demand is illustrated in part on these two pages. Called upon to furnish high grade materials in a hurry and to meet active competition, we immediately filled the requirements of the big operator as well as the individual builder.

People who are best qualified to judge of the present situation predict that there will be no let up in the demand for homes for several years to come. The great decision for the prospective builder at this time is to take advantage of our present facilities and build now. It is easy to see that an increased demand for material not only forces prices upward but reduces the quality of the available supply.

We offer you in this book dependable qualities in house building materials, at prices that are as reasonable as we can make them, and we place at your command the services of an organization that has been perfected to a high degree of efficiency, gained by the experience of many years.

Street view of Sears, Roebuck and Co. houses at Carlinville, Ill.

A few of many Sears, Roebuck and Co. houses at Akron, Ohio.

Street view of Sears, Roebuck and Co. houses at Carlinville, Ill.

Sears, Roebuck and Co. houses at Wood River, Ill., sold to the Standard Oil Company.

SEARS, ROEBUCK AND CO. Page 7

featuring Carlinville's Historic Standard Addition – These would appear in many more issues.

ꟷ **May 21, 1919**, the **Carlinville Democrat** reported, *"The great building boom in Carlinville is so constant that our people are becoming accustomed to it. Not so with visitors. Even the people riding through on the trains look out the windows and remark about the boom that Carlinville is experiencing. The pleasant and gratifying thing about it is that it is permanent and not fleeting."* They were wrong about that statement; it was not permanent.

Carlinville's rapidly growing population is indicative of the great boom that is on here. Every citizen owes it to himself and his community to join in the general movement to boost. The knocker has no place in any community, particularly one such as Carlinville now is. Every legitimate enterprise should be encouraged. Manufacturing industries should now be encouraged to come here. With the power and prestige of the Standard Oil company as a central force to rally around, many other lessor concerns may be brought here. A greater Carlinville means more business, more money and greater prosperity.

Note: There were claims that Carlinville's population doubled as a result of Standard's endeavors. Some have even said that the population of Carlinville doubled over night. Neither of these are true. Carlinville's population did increase considerably, but no ways near double.

Our Schools Are Crowded.

The enrollment in the grades of the schools of Carlinville is 216 more than it was one year ago. This does not give a very good idea of the growth of the city because so many men are working here and boarding, since they are unable to obtain homes in which to live. They probably will bring their families as soon as homes can be provided. There are now in process of construction more than 200 homes. We can reasonably expect many more to be built during the coming spring and summer. Making very conservative estimates as to the number of school children to each house gives us a large number to be taken care by the school next September. The schools are very badly crowded now. What will be the condition when all of these new houses are completed and occupied? Many rooms of the school have an enrollment of nearly sixty pupils. Nearly all the rooms have too many now.

Carlinvlle does not want to have children running the streets because there is no room in the schools for them. This will be the condition next year unless everything is done that is possible to provide more room. It will be impossible to provide room fast enough to prevent a crowded condition, but it will be possible to provide room so that all the children may be able to go to school.

ꟷ Homes at Schoper were finished and occupied in a short time. It seemed that Standard Oil's dream for Carlinville and Carlinville's dream of Standard Oil were both coming true.

*Below is the 1000 block of Sumner Street in Standard Addition looking north in **1924**.*

Everything was going according to plan and Standard was fulfilling their promises. Miners and their families were moving to Carlinville with the promise of full time work at one of the mines and the "American Dream" of owning a home was within their reaches.

Even nearby Nilwood, IL would see a Sears Home built in their town as seen in the photo to the right.

ꟷ The **July 30, 1919**, **Carlinville Democrat** ran an article pleading for a *"New Community High School."* It stated, *"As everyone knows, the activities of Standard Oil have caused a rapid growth in Carlinville. The entire community, as well as Carlinville has been immensely benefited by these activities, therefore the community, and this includes Standard Oil, should in justice assist in building a high school, which the entire community will continue to use as in*

The Schoper Mine site with completed houses.

times past. Then Carlinville will have all it can do to take care of the grade school children." The article went on to further state what would happen if a new high school were not built, *"First, Carlinville, in the near future will be unable, on account of the lack of room, to maintain a community high school. Carlinville will have to bar out of town pupils and have a high school for the district alone. This will be deplorable and cause a great inconvenience to the people of the community."* Then the paper went on to say, *"It is now up to you. Shall that condition prevail in Carlinville? Election August 2nd, go to the polls and vote for the community high school."* Needless to say, it was voted on and passed and a new community high school was built.

Carlinville Community High School **1919.**

ꟹ Carlinville was indeed a thriving community. An article about the ever thriving Carlinville ran in one of the state papers on **October 8, 1919:** *"I have just returned from a trip through central Illinois. During the past years some remarkable change has taken place in the state. In Macoupin County, my old home, land values have increased enormously. Carlinville, the county seat, has become a thriving city. New buildings are going up at a great rate and the pockets of real estate dealers are being relined to contain the showers of blessings brought on by the boom. The immediate cause of the boom is the interest Standard Oil has been taking in the coal fields in the vicinity. Whole farms have given place to development activities, new railroads have been put through, several additional town sites have been laid out and shaft structures and smokestacks dot the prairies."*

ꟹ **November 26, 1919,** Oscar Hewitt of the Chicago Tribune, said, *"This town has a reputation for having the best miners houses in Illinois."* John Black, a union official of West Frankfort, in the southern part of the state, referred to them as *"mansions." "They are so good that Fred M. Jewell, Blackburn College instructor, has purchased one and moved into it. Standard Oil two years ago, purchased the mine, which has been worked here for 30-years. Then it sank another and is developing a 3rd at Schoper, 8-miles northeast. All coal is shipped to Whiting, IN or to Wood River near Alton, IL where it is used for oil refining. The mines,* *therefore, work every day possible. It is the management's intention to work 298 days a year, while many other mines average less than 200."* Frank Fisher, general manager, said, *"Standard Oil decided that men with good, steady incomes should have the opportunity of buying good, modern homes. The company, therefore, erected 156 houses. They were completed within the last 3 weeks. Since they have all been offered for sale, 24 have been disposed of and the manager of the houses, Charles Fitzgerald, said, 'We will sell them all before Christmas. We are offering them all at $1,000.00 less than the cost of reproduction. We contracted for them some time ago and the erection started* ***June of 1918****.'"*

"The company purchased a 40-acre wheat field and the government would not permit destruction of the crop. On the 1st house we were erecting the studding while harvesters were shucking wheat 20-yards away.

42 are 5-room houses and 104 have 6-rooms. Each is on a lot approximately 47 x 144-feet. All of the houses are 2-story frame with asphaltic roof, shaped and colored to resemble slate shingles. The roof carries a guarantee of 15-years. Each house has a front porch, lights, hot air furnace heat, and some have rear porches and a few garages. All have city water, electric porch light, shades and are plastered. Each has a tiled bathroom floor with a skirt bottom tub, medicine case, ornamental washbowl, towel rack and other trimmings. Some are semi-bungalow type, others are house shaped. One has a plain gable roof, another a 4-gable, and another has a protruding roof decorated with a dormer window or two. The floors are stained and varnished yellow pine, the kitchen floor looks like maple. Each has a cement basement, a pantry and a clothes closet for each sleeping room. In the spring, the streets will be paved with asphalt. Trees and shrubs have been planted on each lot. The prices of the houses range from $3,370.00 to $4,000.00. The purchaser pays 1% a month, and in addition, taxes, fire insurance and 6% on the unpaid portion of the cost. A house that sells for $3,500.00 will require $35.00 per month until paid for."

Standard Oil had come to Carlinville in search of coal and they found it.

Standard Oil had succeeded in securing more than adequate housing for their miners. They could now keep good quality laborers at work in their mines.

The town of Carlinville was prospering and booming.

It seemed as though God had smiled down on Carlinville and blessed it. But would it last forever?

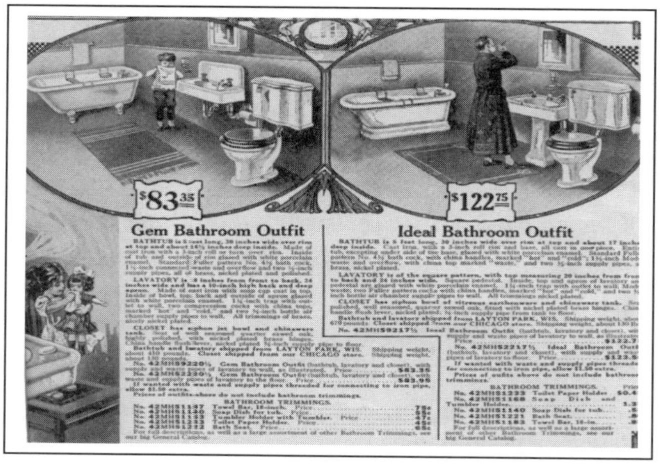

Almost all of the houses in Standard Addition featured the skirted bathtubs in the bathrooms, and all of the bathrooms were upstairs.

STANDARD ADDITION, CARLINVILLE, ILL.

Carlinville, Ill.—No. 1—Two of the most unpopular buildings of Standard Addition are shown here. They are school houses. Nos. 2 and 3—If Autumn comes, these children will leave the playground for the three R's and school. The boys, especially, hope that Autumn won't come. No. 4—Here are some of the residents of Rice street of Standard Addition, where live many of the S. O. Co. mine workers and their families.

A page from the Stanolind Record – Standard Oil's employee newsletter – in the **September 1923** *issue.*

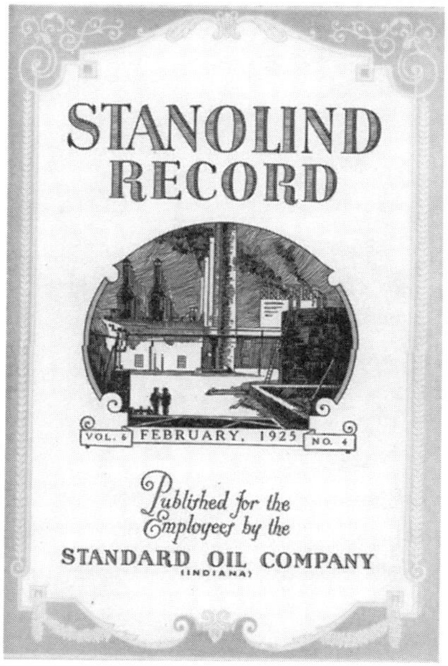

More Standard Addition Photos.

Chapter VI - 1

"You Can Build It"

You Can Build It

☙ Each House Kit Came With A 75-Page Instruction Manual ❧

❧ The Sears Modern Homes program prided itself on the fact that by using their step-by-step instruction method and pre-cut materials, even the most inept of carpenters could build their own home.

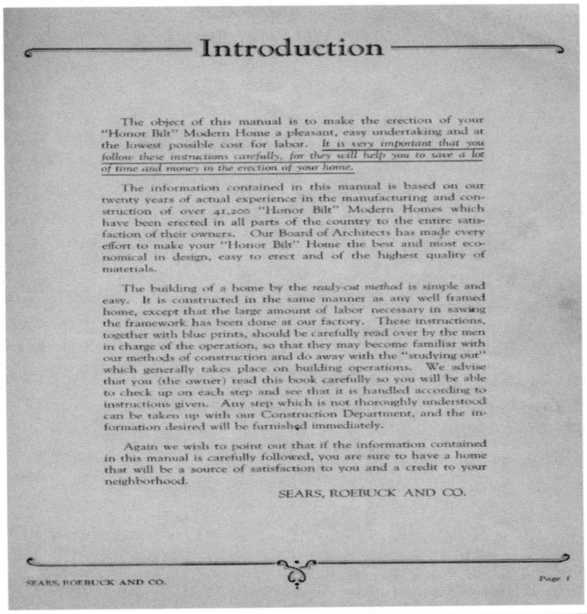

Introduction

The object of this manual is to make the erection of your "Honor Bilt" Modern Home a pleasant, easy undertaking and at the lowest possible cost for labor. *It is very important that you follow these instructions carefully, for they will help you to save a lot of time and money in the erection of your home.*

The information contained in this manual is based on our twenty years of actual experience in the manufacturing and construction of over 41,200 "Honor Bilt" Modern Homes which have been erected in all parts of the country to the entire satisfaction of their owners. Our Board of Architects has made every effort to make your "Honor Bilt" Home the best and most economical in design, easy to erect and of the highest quality of materials.

The building of a home by the *ready-cut method* is simple and easy. It is constructed in the same manner as any well framed home, except that the large amount of labor necessary in sawing the framework has been done at our factory. These instructions, together with blue prints, should be carefully read over by the men in charge of the operation, so that they may become familiar with our methods of construction and do away with the "studying out" which generally takes place on building operations. We advise that you (the owner) read this book carefully so you will be able to check up on each step and see that it is handled according to instructions given. Any step which is not thoroughly understood can be taken up with our Construction Department, and the information desired will be furnished immediately.

Again we wish to point out that if the information contained in this manual is carefully followed, you are sure to have a home that will be a source of satisfaction to you and a credit to your neighborhood.

SEARS, ROEBUCK AND CO.

SEARS, ROEBUCK AND CO. Page 1

They included a 75-page instruction manual with every pre-cut home they sold. This manual entitled *"HOW TO BUILD – Your Ready Cut Honor Bilt Home,"* contained complete start to finish instructions in step-by-step detail of how to build your home from the sorting of the materials when they arrived to staining and painting the interiors.

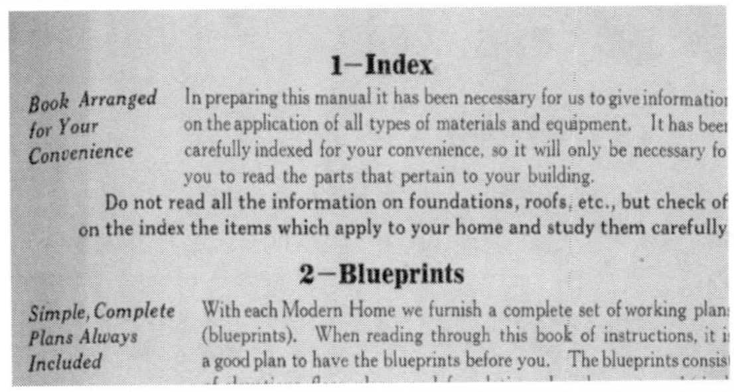

1—Index

Book Arranged for Your Convenience In preparing this manual it has been necessary for us to give informatior on the application of all types of materials and equipment. It has bee carefully indexed for your convenience, so it will only be necessary fo you to read the parts that pertain to your building.

Do not read all the information on foundations, roofs, etc., but check of on the index the items which apply to your home and study them carefully

2—Blueprints

Simple, Complete Plans Always Included With each Modern Home we furnish a complete set of working plan: (blueprints). When reading through this book of instructions, it i a good plan to have the blueprints before you. The blueprints consis

❧ Page one explained to would be home builders, the step-by-step and chapter-by-chapter method of how to build their homes.

❧ Chapter **1**, - "The Index", listed all items in alphabetical order rather than numerical order. This was meant to be used as a checklist for items pertaining to each individual home.

❧ Chapter **2** - contained information about the "Blueprints."

❧ **3** - "Shipment" of your pre-cut Sears House.

❧ **4** - "Unloading and Sorting Material."

❧ **5** - "Protecting the Material."

❧ **6** - "Checking the Material."

❧ **7** - "Laying Out Building Lines."

❧ **8** - "Excavating."

❧ **9** - "Footings."

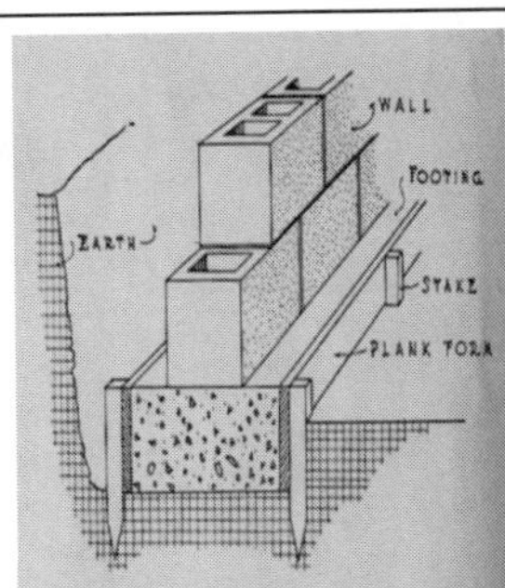

Figure 5. Concrete footings held in place with

Brick Foundation

prepare a footing which will be of good propor-

Figure 20. Brick Bonds.

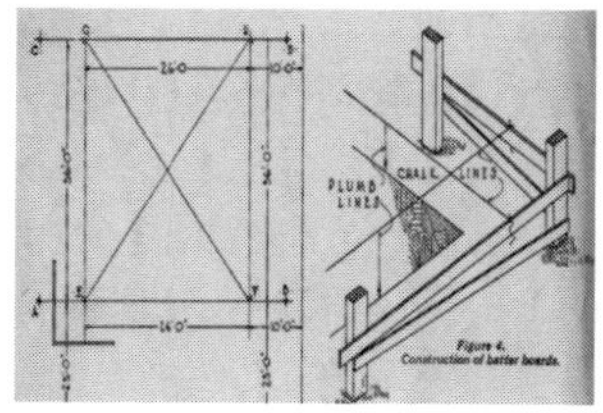

Figure 4. Construction of batter boards.

ꝏ 10 - "Drain Tile."

ꝏ 11 - "Concrete."

ꝏ 12 - "Foundations."

ꝏ 13 - "Concrete Foundations."

ꝏ 14 - "Concrete Block Foundations" – *the type of foundations used in Standard Addition.*

ꝏ 15 - "Concrete Floors."

ꝏ 16 - "Porch Concrete Floors and Steps."

ꝏ 17 - "Brick Work."

ꝏ 18 - "Laying a Brick Foundation."

ꝏ 19 - "Estimating Quantities of Brick."

ꝏ 20 - "Mortar."

ꝏ 21 - "Brick Veneer."

ꝏ 22 - "Chimneys."

ꝏ 23 – "Fireplaces."

ꝏ 24 - "Stucco."

ꝏ 24A - "Finishes."

ꝏ 25 - "Rough Hardware."

ꝏ 26 - "Nails."

ꝏ 27 - "Building Paper."

ꝏ 28 - "Foundation, Piers & Girders."

ꝏ 29 - "Wall Plates."

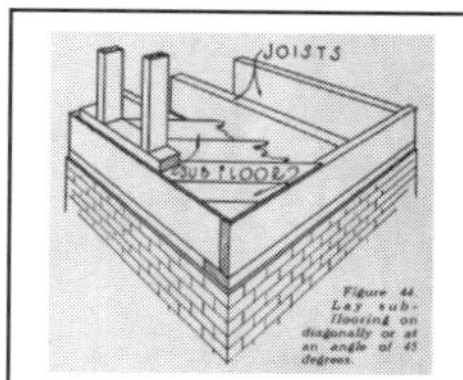

Figure 44. Lay subflooring on diagonally or at an angle of 45 degrees.

Note: Sears did not have pre-determined quantities of items that were automatically sent out with the houses. Everything was based on tables similar to the nail amounts, which were determined according to the size of house that was purchased and based on the table to the right.

14—Concrete Block Foundation

Concrete Blocks Have Many Advantages A great advantage in concrete block walls is the use of forms, which results in hollow walls, that is, blocks with one or more hollow spaces, or blocks so constructed that their combination in a wall produces hollow spaces in it. These dead air spaces make the wall more impervious to moisture and act as an insulation against heat and cold, also saving material.

A dry basement is a positive necessity of good construction. For ordinary well drained soils, the most important precaution in building the concrete block walls below grade is to see that joints are well filled with cement mortar and well pointed. If the sub-soil does not have good drainage, a line of drain tile placed entirely around the outside of the footings and connected to a suitable outlet generally can be relied on to carry off excess water. Care should be exercised to excavate to proper grade so that there will be a uniform slope to the tile with no low places. The excavation above the tile should be filled to a depth of 1 or 2 feet with gravel, cinders or some other material of porous nature through which water can seep easily. It is always a good plan to place pieces of tar paper over the tile joints so they will not become filled with mortar. If for any reason it is impossible to run a line of tile around the outside, the tile may be placed on the inside of the footings and below the lowest point of the cement floor.

If the soil is very retentive of moisture or the water level is likely to rise above the footings, the usual treatment then is to apply to the exterior wall two or three coats of Medusa Waterproofed Gray Portland Cement mortar mixed in the proportion of 1 sack of cement to 2 cubic feet of clean, well graded sand.

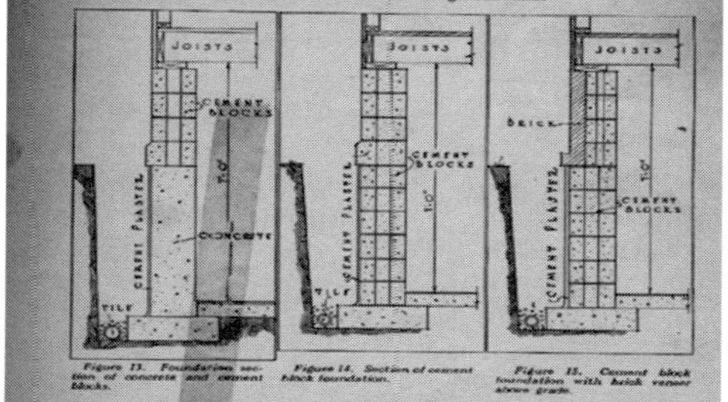

Figure 13. Foundation section of concrete and cement blocks. Figure 14. Section of cement block foundation. Figure 15. Cement block foundation with brick veneer above grade.

SEARS, ROEBUCK AND CO. Page 15

26—Nails

Use Only Correct Nails for Each Purpose All the nails which have been furnished for your building are for a certain purpose and the following table should be used. Do not leave your nails out over night, for if it should rain they will become rusty and hard to handle.

5 pounds 3d galvanized nails for each 1,000 wood shingles.
6½ pounds 3d blue fine nails for each 1,000 lath.
18 pounds 6d box nails for each 1,000 feet 4-inch bevel siding.
14 pounds 6d box nails for each 1,000 feet 6-inch bevel siding.
30 pounds 8d common nails for each 1,000 feet drop siding.
30 pounds 8d flooring nails for each 1,000 feet flooring.
20 pounds 4d casing nails for each 1,000 feet ⅜-inch flooring.
15 pounds 6d finishing nails for each 1,000 feet ⅝-inch ceiling.
25 pounds 8d common nails for each 1,000 feet ¾-inch partition.
16d common nails for heavy framing work.
10d common nails for toe-nailing and studs.
8d common nails for rough 1-inch lumber.
20 pounds 8d box nails for each 1,000 feet ½x8-inch siding.
20 pounds 3d common nails for each 1,000 feet plaster board.
5 pounds roofing nails for each square O. S. S, shingles.
10d casing nails to apply 1⅛-inch outside trim and window and door frames.
8d casing nails for window frames, porch work, etc.
1-inch brads for small stops and molds.
6d finishing and 4d finishing for small mold members.

SEARS, ROEBUCK AND CO. Page 33

35—Erecting Frame Outside Walls

Blueprints Show Material Required You are now ready to erect the frame for the outside walls. Carefully study the blueprints; note the walls are so constructed as to interlock at the corners. The walls consist of floor plates or the plates which are nailed to the subfloor of the building, upright studs, window and door headers and two sets of wall plates or plates which cap the upright studs. (Study Figure 45.) Time can be saved by having one man start to lay out the walls as soon as sufficient subflooring has been laid across the joists. The blueprints show a bill of material of what is required for each section.

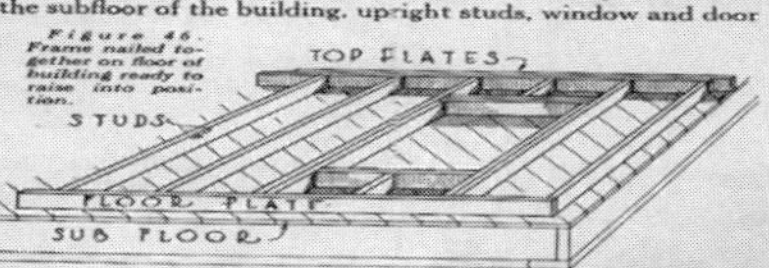

Figure 46. Frame nailed together on floor of building ready to raise into position.

To lay out the walls, first lay the floor plates on top of the subflooring and lay one set of top wall plates alongside of them; now space off with a steel square, the location of the upright studs in the wall. Space the window openings first. Lay these plates on floor of building and nail the upright studs in between them; then nail in place the window and door headers. Note Figure 46, which shows all the frame for one wall nailed together ready to be raised into position; raise it into position and spike to floor of building. Carefully study Figure 47, which shows wall raised and held in place with supports. Square and plumb the wall, and nail in place sufficient temporary supports until the remaining walls and partitions have been erected. Use the second set of top wall plates to help line up the wall.

You will note that temporary supports are tacked on upright studs before they are raised in place. In this manner the wall can be raised with less men. The opposite end of supports is tacked to a small block on top of the subflooring and securely nailed after wall has been squared up.

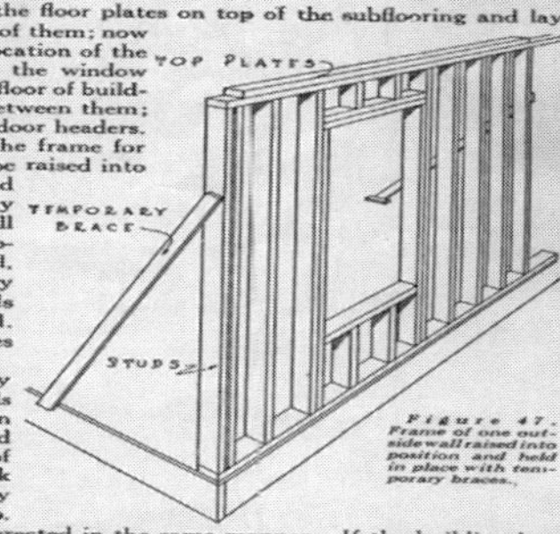

Figure 47. Frame of one outside wall raised into position and held in place with temporary braces.

All the outside walls should be erected in the same manner. If the building has two stories, the second story outside walls will be erected on top of the subflooring of the second floor in the same manner as the first story walls. At the corners where the wall plates intersect, they should be carefully spiked together.

SEARS, ROEBUCK AND CO. Page 39

- **30** - "Built Up Girder."
- **31** - "Framing – Use of Steel Square."
- **32** – "Laying Joists First Floor."
- **33** - "Wood Bridging."
- **34** - "First Floor Subflooring."
- **35** - "Erecting Frame Outside Walls."
- **36** - "Partitions."
- **37** - "Sheathing Outside Walls."
- **38** - "Second Floor Joists."
- **39** - "Ceiling Joists."
- **40** - "Gable Roof" – *one style in Standard Addition – The Warrenton - has this type of roof.*
- **41** - **"**Hip Roof" – *two styles in Standard Addition – The Gladstone and Langston - have this roof type.*
- **42** - "Collar Beams."
- **43** - "Gable Studding and Sheathing."
- **44** - "Roof Sheathing."
- **45** - "Dormers & Porches."
- **46** - "Eave Construction."
- **47** - "Wood Shingles."
- **48** - "Slate Surface Shingles."
- **49** - "Roll Roofing."
- **51** - "Outside Finish."
- **52** - "Outside Window and Door Frames."

Note: This book was designed so that even a person with the most limited carpentry skills could follow these simple instructions and build their home – How ingenious!!

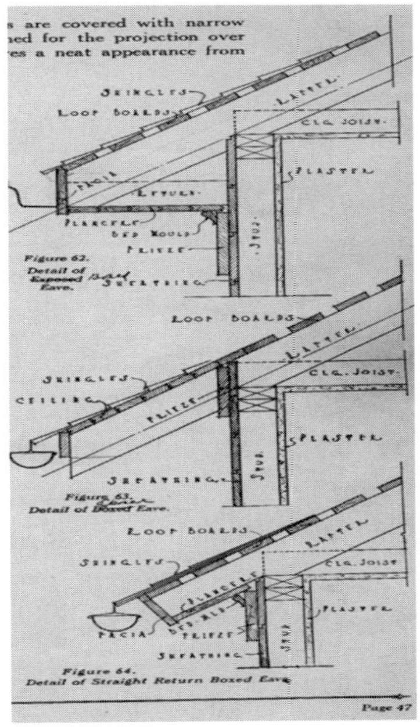

Figure 62. Detail of Exposed Eave.

Figure 63. Detail of Boxed Eave.

Figure 64. Detail of Straight Return Boxed Eave.

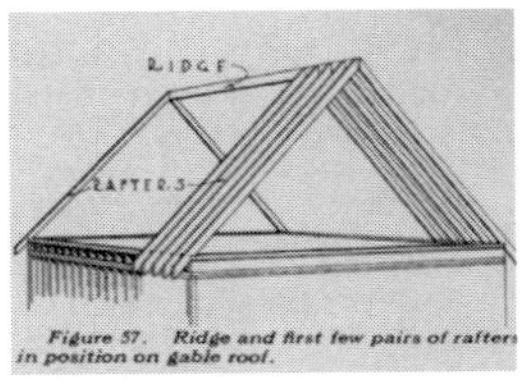

Figure 57. Ridge and first few pairs of rafters in position on gable roof.

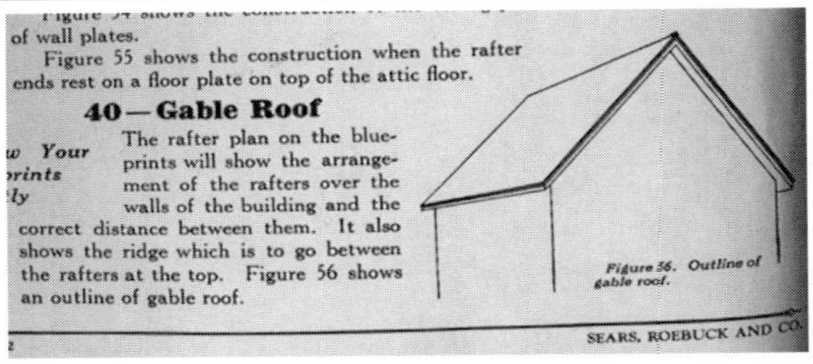

of wall plates.

Figure 55 shows the construction when the rafter ends rest on a floor plate on top of the attic floor.

40—Gable Roof

...w Your ...rints ...ly

The rafter plan on the blueprints will show the arrangement of the rafters over the walls of the building and the correct distance between them. It also shows the ridge which is to go between the rafters at the top. Figure 56 shows an outline of gable roof.

Figure 56. Outline of gable roof.

SEARS, ROEBUCK AND CO.

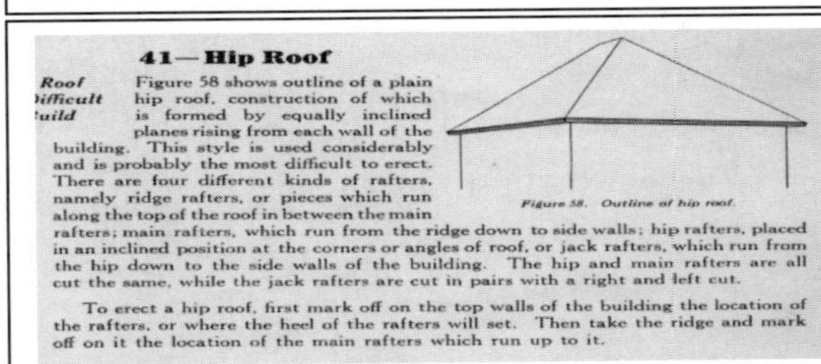

41—Hip Roof

Roof ...ifficult ...uild

Figure 58 shows outline of a plain hip roof, construction of which is formed by equally inclined planes rising from each wall of the building. This style is used considerably and is probably the most difficult to erect. There are four different kinds of rafters, namely ridge rafters, or pieces which run along the top of the roof in between the main rafters; main rafters, which run from the ridge down to side walls; hip rafters, placed in an inclined position at the corners or angles of roof, or jack rafters, which run from the hip down to the side walls of the building. The hip and main rafters are all cut the same, while the jack rafters are cut in pairs with a right and left cut.

To erect a hip roof, first mark off on the top walls of the building the location of the rafters, or where the heel of the rafters will set. Then take the ridge and mark off on it the location of the main rafters which run up to it.

Figure 58. Outline of hip roof.

Note: Sears made a mistake in printing their manual and left out chapter 50; the manual went from 49 to 51.

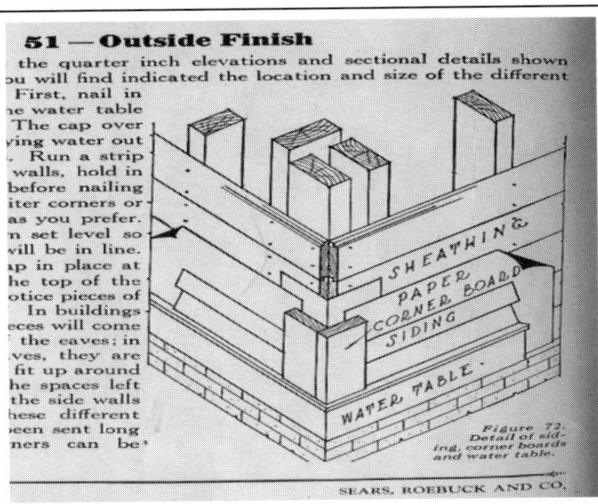

51 —Outside Finish

...the quarter inch elevations and sectional details shown ...ou will find indicated the location and size of the different ... First, nail in ...ne water table ... The cap over ...ying water out ... Run a strip ... walls, hold in ...before nailing ...iter corners or ...as you prefer. ...n set level so ...vill be in line. ...ap in place at ...he top of the ...otice pieces of ... In buildings ...eces will come ... the eaves; in ...ves, they are ...fit up around ...he spaces left ...the side walls ...hese different ...een sent long ...ners can be...

Figure 72. Detail of siding, corner boards and water table.

SEARS, ROEBUCK AND CO.

ഇ **53** - "Siding."

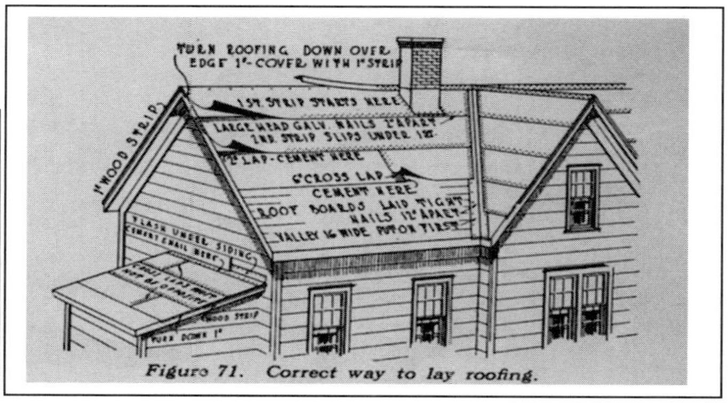

Figure 71. Correct way to lay roofing.

ഇ **54** - "Shingled Side Walls."

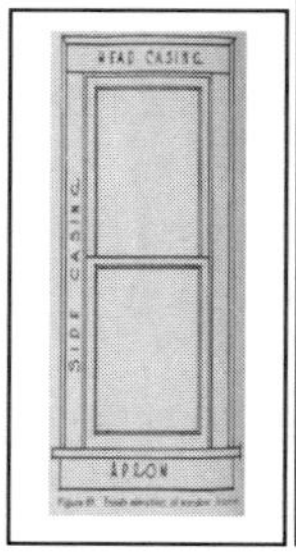

ഇ **55** - "Gutters."

ഇ **56** - "Fixture Supports."

ഇ **57** - "Plasterboard."

ഇ **59** - "Instructions for Mixing and Applying Plaster Finish."

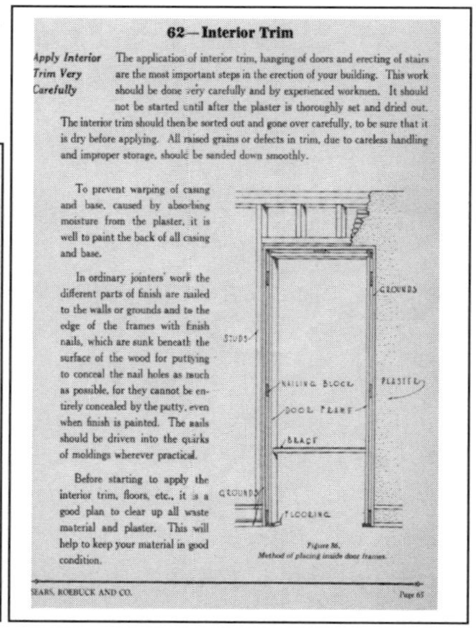

62—Interior Trim

Apply Interior Trim Very Carefully The application of interior trim, hanging of doors and erecting of stairs are the most important steps in the erection of your building. This work should be done very carefully and by experienced workmen. It should not be started until after the plaster is thoroughly set and dried out. The interior trim should then be sorted out and gone over carefully, to be sure that it is dry before applying. All raised grains or defects in trim, due to careless handling and improper storage, should be sanded down smoothly.

To prevent warping of casing and base, caused by absorbing moisture from the plaster, it is well to paint the back of all casing and base.

In ordinary jointers' work the different parts of finish are nailed to the walls or grounds and to the edge of the frames with finish nails, which are sunk beneath the surface of the wood for puttying to conceal the nail holes as much as possible, for they cannot be entirely concealed by the putty, even when finish is painted. The nails should be driven into the quirks of moldings wherever practical.

Before starting to apply the interior trim, floors, etc., it is a good plan to clear up all waste material and plaster. This will help to keep your material in good condition.

Figure 86. Method of placing inside door frames.

SEARS, ROEBUCK AND CO. Page 65

ഇ **60** - "Plaster Difficulties and How To Correct Them."

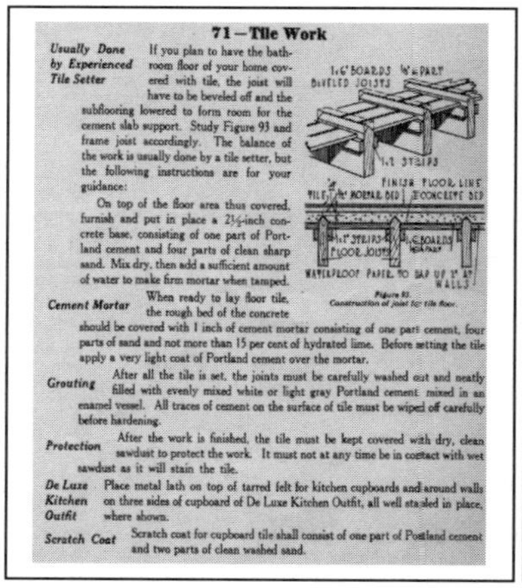

71—Tile Work

Usually Done by Experienced Tile Setter If you plan to have the bathroom floor of your home covered with tile, the joist will have to be beveled off and the subflooring lowered to form room for the cement slab support. Study Figure 93 and frame joist accordingly. The balance of the work is usually done by a tile setter, but the following instructions are for your guidance:

On top of the floor area thus covered, furnish and put in place a 2½-inch concrete base, consisting of one part of Portland cement and four parts of clean sharp sand. Mix dry, then add a sufficient amount of water to make firm mortar when tamped.

Cement Mortar When ready to lay floor tile, the rough bed of the concrete should be covered with 1 inch of cement mortar consisting of one part cement, four parts of sand and not more than 15 per cent of hydrated lime. Before setting the tile apply a very light coat of Portland cement over the mortar.

Grouting After all the tile is set, the joints must be carefully washed out and neatly filled with evenly mixed white or light gray Portland cement mixed in an enamel vessel. All traces of cement on the surface of tile must be wiped off carefully before hardening.

Protection After the work is finished, the tile must be kept covered with dry, clean sawdust to protect the work. It must not at any time be in contact with wet sawdust as it will stain the tile.

De Luxe Kitchen Outfit Place metal lath on top of tarred felt for kitchen cupboards and around walls on three sides of cupboard of De Luxe Kitchen Outfit, all well stapled in place, where shown.

Scratch Coat Scratch coat for cupboard tile shall consist of one part of Portland cement and two parts of clean washed sand.

Figure 93. Construction of joist for tile floor.

ഇ **61** - "Hanging Sash."

ഇ **62** - "Interior Trim."

ഇ **63** - "Outside Door Frames & Trim."

ഇ **64** - "Inside Door Frames."

ഇ **65** - "Window Trim."

ഇ **66** - "Laying Finish Flooring."

Note: There was a blank "Plot Plan Sketch" area in the back of the book to use in planning your home or garage position – each square was equal to 2-feet. This was followed by several blank pages to use for figuring. By the time you completed the book, your home should have been finished as well. The instruction book was bound in black leather and Sears embossed your name on it in gold.

ഇ **67** - "Hardwood Floors."

ഇ **68** - "Base Molding."

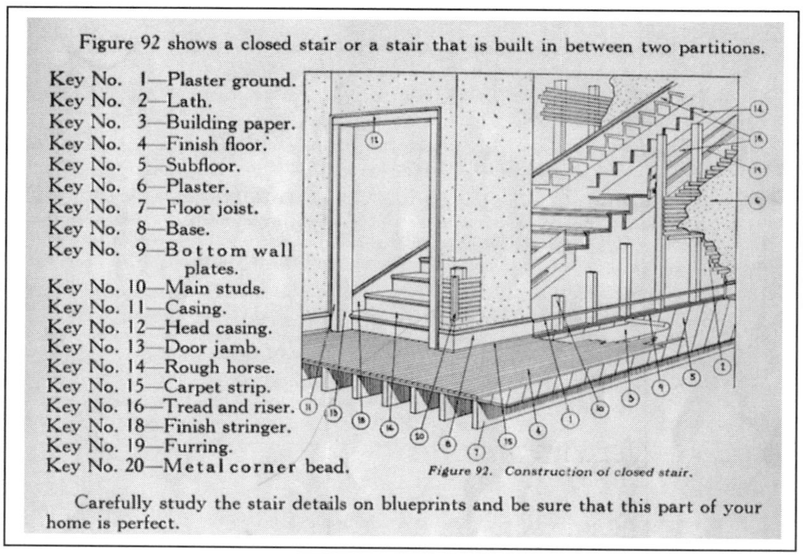

Figure 92 shows a closed stair or a stair that is built in between two partitions.

Key No. 1—Plaster ground.
Key No. 2—Lath.
Key No. 3—Building paper.
Key No. 4—Finish floor.
Key No. 5—Subfloor.
Key No. 6—Plaster.
Key No. 7—Floor joist.
Key No. 8—Base.
Key No. 9—Bottom wall plates.
Key No. 10—Main studs.
Key No. 11—Casing.
Key No. 12—Head casing.
Key No. 13—Door jamb.
Key No. 14—Rough horse.
Key No. 15—Carpet strip.
Key No. 16—Tread and riser.
Key No. 18—Finish stringer.
Key No. 19—Furring.
Key No. 20—Metal corner bead.

Figure 92. Construction of closed stair.

Carefully study the stair details on blueprints and be sure that this part of your home is perfect.

ഇ **69** - "Stairs."

ഇ **70** - "Finish Hardware."

ഇ **71** - "Tile Work."

ഇ **72** - "Painting."

ഇ **73** - "Shingle Stain."

ഇ **74** – "Interior Painting."

Chapter VII

“Time Went On”

Time Went On

ꟷ By early **November of 1919,** all of the Standard Addition houses were occupied. Happy miners with their families were a common site in Carlinville. Standard Addition, as well as its occupants became a part of the Carlinville community. The Standard Addition homeowners were working hard in the Schoper and Berry Mines and coming home to beautiful homes at the end of the day. Clubs were formed as were baseball teams, soccer teams, organizations and more. Time went on.

What follows is life in Carlinville, Standard Addition and the mines November of 1919 on…

A Coal Strike

"The Coal Strike Is On," was the headline of every newspaper in the U.S.A. as well as the **Carlinville Democrat's November 5, 1919** issue. It reported that, *"Hundreds of mines were closed and more than 400,000 miners were idle. Nearly all of the coal mines in the country are closed as a result of the strike of miners, which took effect Saturday."* Reports showed that Illinois had the most men idle, being more than 90,000.

Agreement Reached

In **December 1919**, the coal strike ended. Agreements had been reached between the government and mining officials. A commission appointed by President Wilson was to investigate the mining industry and report within 3 or 4 months. The strike had lasted 6 weeks.

The **December 1919,** Stanolind Record (Standard Oil Company's employee newsletter) issue said that ,*"Standard Oil Mine management officials are engaged in a continual struggle to obtain railroad cars to ship the coal. How unending and difficult this struggle for coal cars is can only be understood by spending a few days in the managers office."*

A New Community High School

January 14, 1920, the **Carlinville Democrat** about the coming election. *"The election is Sunday, January 24th and polls will be open from 9 a.m. until 8 p.m. The voting place is the northeast corner of the courthouse basement. The election of the site to build the new community high school, which was voted in last August, is to be determined by this vote. Some sections have filed petitions to withdraw from the district – two of theses are Schoper and the big power plant. This district surely realizes the importance of this election –go to the polls and vote."*

New Management at the Mines

On **January 1920,** Frank Fisher, who was manager of Standard Oil Mines for Carlinville, Berry and Schoper, resigned. C.W. Clark, who had been with Standard for a long time and in Carlinville for months, was appointed to fill the position. Ben Firth, superintendent of the mines, resigned; and Thomas W. Paul was appointed to succeed him. John McMillan was appointed safety inspector for Mine 1-A – the old Carlinville Mine and 1-B – the Berry shaft.

☙ Berry Mine Gets Electricity ❧

❧ Late **January 1920,** Standard installed automatic pump cages at the Carlinville Mine and concreted the shaft at Schoper. High-tension wires from Schoper to Berry were brought in and in a short time they carried electricity for operating the machinery at Berry.

☙ Carlinville's Population Increases to 5,271 ❧

❧ **April 21, 1920,** the Carlinville Democrat reported that the population in Carlinville was now at 5,271. The paper stated, *"Increase during the last 10-years was 44.1 percent. The population of Carlinville was 3,616 in **1910** – a wonderful increase in 10-years. This was due in part to Standard Oil, who have spent several millions of in the last 3-years in Carlinville."*

☙ Carlinville Standard Oil Co. Baseball Team "The Stanolinds" ❧

❧ **May 19, 1920,** *"Next Sunday at the baseball park in Standard Addition, Myers Brothers team of Springfield, will play the Stanolinds of Carlinville."* The Stanolinds ball team was named after Standard's new cheaper gas, introduced to bring their gasoline sales out of the red. This gasoline was named "Stanolind Blue" by Standard, as it had a blue color added to it to distinguish it from different gasoline products.

Right is "Stanolind A" in the lock at Alton, IL. This towboat was capable of pushing 6 barges loaded with 2,400,000 gallons of petroleum against Mississippi River currents.

Standard Oil's private airplane called the "Stanolind."

Special Thanks to Thomas R. Giddens, Jackson A. Giddens & Judy (Giddens) White for use of Stanolind images from their father, Paul H. Giddens' book "Standard Oil Company of Indiana – Oil Pioneer of the Midwest."

☙ Illegal Switchmen's Strike ❧

❧ The **May 1920,** Stanolind Record said that, *"Miners at mines in Carlinville suffered the fate of the innocent bystander in connection with the illegal switchmen's strike. Because the coal could not be moved it was impossible to operate the mines. On one occasion, a full trainload of cars remained on the track for 3 days before the coal was moved."*

☙ Macoupin County Population Increases ❧

❧ **June 9, 1920,** Macoupin County's population was at 57, 274.

☙ Accident at Schoper Mine, Death at Berry Mine ❧

❧ **June 9, 1920,** the local paper reported an accident at Schoper: *"Saturday while at work at*

Schoper, Ralph Hall had two fingers on one hand badly crushed and they had to be amputated." The same paper also reported that a miner had been killed at Berry Mine: *"Battista Marchiando was killed by a fall of rock in Berry Mine on Saturday June 5th."*

Train Wreck at Sheep's' Crossing North of Carlinville

On **June 23, 1920,** there was a train wreck at Sheep's Crossing north of Carlinville. At 3:00 a.m., a number of loaded coal cars were switched out from Berry onto the main line. A fast southbound freight train ran into the cars and traffic was tied up until noon. The wooden cars were badly damaged but the steel cars came out okay. The engineer received bruising.

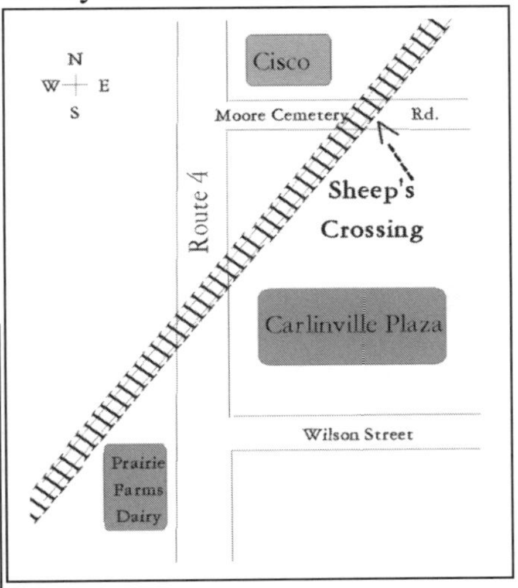

Standard City Grocery Store Robbed

July 7, 1920, the grocery store owned by J.J. Shutt of Standard City was held up. Robbers forced the clerk, at the muzzle of revolvers, to enter the icebox and robbed the cash drawer of $450. No trace of the robbers was found.

Left - A photo of the old Standard City Grocery Store.

Death at Schoper

On **October 27, 1920,** there was a death at Schoper. Roland Gamber, a machine runner, was killed by a fall of slate.

THE CHAMPION CARLINVILLE BLACK DIAMONDS

Here are the Carlinville Black Diamonds, leading exponents in the Standard Oil family of the well-known game of soccer. They have twice defeated the Wood River Polarines, though the oil men gave them a battle. Left to right: Simpson (manager), Carrol, McWilliams, T. Brown, McKeegan, Clark, Benjamin, Palmeto, Jack, McGuire, Hibbard (captain), Brown.

Standard's Soccer Team – 1921.

Standard Oil Company Purchased Farm

November 3, 1920, Standard purchased the 80-acre farm of A. H. Bell located just north of Berry Mine. The tract of land joined Standard Oil Company's property on the north.

Carlinville Standard Oil Has Soccer Team

On **December 1, 1920**, Standard Oil Soccer team of Carlinville defeated the Wood River Refinery team on Wood River Refinery grounds by a score of 2 to 1.

Engineer of Mines Robbed

December 8, 1920, H. M. Fenton, engineer of the mines, was robbed. One robber pointed a revolver at Fenton and demanded his money. He gave them all he had, which was $1.00.

~ Still Captured in Standard Addition ~

~ **December 22, 1920,** the Carlinville Democrat said a still had been captured in Carlinville. *"Thursday at 7:00 p.m., the sheriff, his men and the local police force, assisted by officers from Springfield, made a raid in the Standard Oil Addition at the home of Frank Jacklitch. They found a still and white mule. Jacklitch was arrested and taken to the county jail and locked up."*

~ Paving Ordinance Gets Objections ~

~ On **December 23, 1920**, the paper claimed citizens were objecting to the paving of Charles Street, *"Sometime ago, the City Council adopted an ordinance on the Charles Street Paving. This ordinance regards paving Charles Street running north & south with the southern end on the south side of the block next to the city cemetery. The north end runs through Standard Addition. At the time this ordinance was adopted and before, there were objections by property owners and no one on this street asked for the improvement. A sewer had been put down Charles Street a year before and assessment for this was heavy. To bring this matter to the attention of the court, 50 citizens living along Charles Street, met at the courthouse. A motion was adopted to employ Attys Victor Hemphill & Jesse Peebles to go before the court and fight the assessment roll. The matter was thoroughly discussed. The citizens want it distinctly understood that they are not fighting Standard Oil, but are objecting to paying for improvements along their property about which they were not consulted, but on which they have to pay the 'freight' if it goes through."*

~ Safety Organization Formed ~

~ **December 29, 1920,** Standard called a meeting of its officials for the purpose of installing a safety organization to reduce the amount of minor and major accidents involving the miners in the mines. Ralph Brown was appointed Safety Manager.

~ Carlinville Mine Closed ~

MINE ONE-A, CARLINVILLE, DISCONTINUED

~ **December 29, 1920,** Standard announced that they would close Mine No 1 – the old Carlinville Mine – in **January.** The reason was cited as the expense of hauling coal from the works to the main shaft was too heavy and it was cheaper to take the coal to Berry. Miners from Mine No. 1 were offered jobs at Berry.

~ **January 19, 1921,** Supt. Weller, of the committee to which was referred the matter of the sale of coal under the county farm reported

"The county should not accept the bid of Standard Oil of $50.00 per acre, at this time." Mr. Williamson, representing Standard Oil, withdrew all propositions made by Standard Oil regarding coal under the county farm.

ꟹ **February 2, 1921,** Standard Oil Mothers' Club met with 18 members present. A delightful afternoon was spent. They voted to give $10.00 to the 3 relief funds that were then being raised, and also voted to, each earn and donate $1.00 to some charitable and worthy cause.

ꟹ The **February 1921,** Stanolind Record said that, *"James Scott and Fred Scott of Troy, IL, while visiting their brothers, Alfred and Tom, were so impressed with Carlinville and especially with our beautiful homes in Standard Addition that they decided to make Carlinville their future residence."* This same issue also said, *"We noticed some new faces at 1-B lately. Boys from 1-A we guess."* ***Note: This was talking about the closing of the Carlinville Coal Mine and all of the employees going to Berry to work.*** It was also reported that, *"E. Hawthorne, engineer at Mine 1-B, was severely injured between a load of flat cars and his engine, when he crashed into a string of loaded coal cars."*

ꟹ The **March 1921,** Stanolind Record said, *"From Standard Addition comes the rumor that two of our Schoper men, Harry McWilliams and Elmer Boode, are waging desperate warfare against a gang of crooks who appear to be infesting their neighborhood."* There had been several robberies and vandalism reported in the Addition for the previous month.

Injuries at Berry Mine

ꟹ **April 20, 1921,** two men employed at Berry were injured in accidents. Samuel Baker was hurt in the back and Ted Dauost broke his leg by a fall of slate.

ꟹ The **April 1921,** Stanolind Record stated that Mine 1-B hoisted 725 cars of coal. They called the miners real *"coal getters."*

ꟹ **May 11, 1921,** motor driver, Albert Galli, was killed at Berry.

Standard Oil Company Recognized

ꟹ **May 25, 1921,** the Carlinville Democrat recognized Standard for their building activities in and near Carlinville. *"It has been thought of the men who have the shaping Standard Oil's policies that its workmen are entitled to more than mere shelter. These houses have been built with bath, furnace, electrically lighted, sewered, and the streets will be paved without further cost to the purchasers. Homeownership makes for permanency."*

A View of the Power House and Reservoir at Schoper

Schoper Power House.

Right is the 900 block of Johnson Street looking north – my home is the third house from this end.

Some of the modern, up-to-date homes built for employes. Already there are 156 o these residences built and occupied, or ready for occupancy, in one addition to Carlinville

Left is the Standard Oil Company Office on North Broad Street in Carlinville - this building is located to the north of the Carl's IGA Store and is now home to Mt. Nebo Lodge.

Home Building a Feature of Standard Oil Co.'s Activities in Carlinville

The Handsome and Commodious Office Building on North Broad Street

A View of the Power House and Reservoir at Schoper

Some of the modern, up-to-date homes built for employes. Already there are 156 of these residences built and occupied, or ready for occupancy, in one addition to Carlinville.

THE basic principle of all constructive social orders and organizations is the betterment of the individual and the resulting advancement of humanity. The foundation of all social order, some dangerous experimental theories to the contrary, is the home. To promote the stability of our whole social order should be every right thinking person's endeavor.

The Standard Oil Co. (Ind.) in extending its activities to this community has followed these principles. Its industrial construction and development speaks for its policy. And to enable its employes to share in its experience they have constructed a substantial addition of modern five and six room houses with the conveniences which modern homes demand.

It has been the thought of the men who have the shaping of this Company's policies that its workmen are entitled to more than a mere shelter. These houses have been built with bath, furnace, electrically lighted, sewered, and the streets will be paved without further cost to the purchasers.

Home ownership makes for permanency. No person ever progressed without a definite plan of action. No person ever owned a home without a definite plan and purpose of ownership. The greatest obstacle to home ownership is the means of acquirement. Most of the plans for home ownership have been formulated on the assumption that the person who has not been able to save at least one-tenth to one-third of the cost of their future home have had no definite purpose.

This Company does not hold that theory. It believes that the employer owes the faithful employee a regular and constant employment and that given such employment the majority of men plan for permanency. To enable its employees to overcome all obstacles of the past in home ownership it has employed a considerable sum of capital, without any purpose or possibility of financial profit to the Company, in the construction of this modern addition of homes to the city of Carlinville. It has enabled its employees to have a definite purpose and will of home ownership to realize their desire, and in so doing become a permanent addition to society. It feels that this is necessary to a proper growth of character in the indi-

The half page article that appeared on the front page of the ***March 25, 1921*** **Carlinville Democrat** *praising Standard Oil for their building projects in Carlinville.*

so The **May 1921,** Stanolind Record stated, *"Standard Addition is getting to be a regular jazz center as several player piano's have been sold there."* It went on to say, *"Our Blacksmith, Geo Loehnberg, recently changed his residence from Standard Addition to Schoper. He says he likes it better as he doesn't have to get up before breakfast now to catch the train for work. Several days ago, Mine No.2 made a record hoist of 1,084 tons. Their previous record was 1,024. We hear that Standard Addition will come out of the mud on June 1."* **Note: This meant that the dirt**

streets in Standard Addition were going to be paved. Then the newsletter said, *"John Swanson, our efficient mechanic on the rotary dumps, and his roommate Ike Smith, a teamster, are building a summer home in the beautiful Schoper Creek Valley. It is said that this stream is noted for its scenic beauty."*

ꟷ **June 6, 1921,** the Carlinville City Council met and an ordinance was read *"All property owners owning property adjacent to Charles Street, Seminary Avenue, Rice Street and all streets in Standard Addition to Carlinville, West Street, Plum Street, First South Street and South Broad Street are required to connect to a sewer within a specified time."* On motion, the ordinance was adopted and put into effect.

ꟷ The Stanolinds defeated the Springfield Merchants on **June 26, 1921** by a score of 3 - 0.

ꟷ The **July 1921,** Stanolind Record stated, *"Work is progressing rapidly on the motor barn. The new washhouse and machine shop have been laid out and the pouring of their foundations has almost been completed."* The past Sunday, "*Mayor Daley opened the baseball season in Carlinville by pitching the 1st ball in the game between Gillespie and the Stanolinds of Carlinville. The Stanolinds won 4 – 2."*

Boy Severely Injured

ꟷ On **August 24, 1921,** the **Carlinville Democrat** reported that, *"Harold Ray, son of Mr. and Mrs. Edward Ray, was injured while playing with a fan at the old airshaft of Standard Oil located just northeast of Fieker Elevator. The fan is not in use, and the Ray boy, with a number of other boys, were amusing themselves turning the fan by hand when it started revolving and one of the steel blades hit the boy on the neck inflicting a serious wound. His condition is critical."*

ꟷ The **September 1921,** Stanolind Record stated, *"There have been a couple of accidents due to slate falls: Daniel Thomas has a slight concussion due to a slate fall and Andrew Griffin suffered a painful bruise to his foot due to falling slate. The latest record at Mine No. 2 is 1,350 ton hoisted. The reservoir at the power house is getting to be quite a swimming resort."*

ꟷ In **September 1921,** the hard road to Girard, which connected with the cinder road to Schoper, opened up, this meant that a road from Carlinville to Schoper would be open all winter.

ꟷ The **October 1921,** Stanolind Record claimed, *"There was an exciting incident at Schoper when the rescue team was practicing. It was fixed before hand that Jack Welch, our 225 lb. boss was going to be overcome in the gas chamber while wearing his breathing apparatus. This he did and the first thing that rescuers Eddie Lewis and Paul Burns noticed was that Jack was on his back in an unconscious state. At first they thought he was kidding, but as he continued to lie there they decided to save him, so they turned him on his stomach and drug him out. Jack said next time that he hoped he really was unconscious because they nearly ruined him **– I think we all know what he's talking about by the word 'ruined' –** getting him out."*

ꟷ **December 7, 1921,** Standard Oil furnished cinders and farmers near the 3-mile stretch of road running from Atwater to Schoper put the cinders on the highway. General Manager

Manbeck, Local Manager C. W. Clark and Thomas Williamson represented Standard Oil in a business meeting held in Carlinville.

☙ **December 7, 1921,** Freno Senchetti died at St. John's Hospital in Springfield, IL as the result of an accident in Berry Mine when a large quantity of slate fell, breaking his neck.

❧ Carlinville Has Five Wards ☙

☙ On **December 21, 1921,** it was reported, *"Carlinville now has five wards. The 5th was added Monday during the City Council meeting. The new ward is bound south by Nicholas Street and west by North Broad Street extending east to the city limits. The old wards were left as they were. By unanimous vote the 5th ward was approved."*

Note: This 5th ward encompassed the new tracts of Standard Oil land incorporated into the city limits. It contained Standard Addition.

❧ Big Sleet Storm Covers County ☙

☙ **December 28, 1921,** a sleet storm covered the county. It was said to be, *"One of the worst sleet storms in Illinois. It hit Carlinville and Macoupin County starting slow, then sleet began to form increasing steadily, continuing all day Friday into the night. Saturday the whole out of doors was a mass of ice. Big tree limbs in all parts of Carlinville were bent to the ground. Some limbs came crashing down with their big load of ice. Telephone poles in the county went down and it is estimated that 100's are broken or damaged. There are many broken wires all over. The utility company's poles sustained heavy damage and power service has been crippled. It will be quite a while before power is restored."*

☙ Mr. W.H. Meyers was hard at work paving streets in Standard Addition according to the **October 1921,** Stanolind Record.

☙ **February 1, 1922**, C.F. Brueggamann of Carlinville was at the county home working on machinery in the well. Water was supplied for the county home by a well 102-feet deep. For a few year's, the well had been failing and at times could not furnish water and water was hauled from Carlinville. It is believed when Berry Mine was sunk, the stream, which supplied the well, was tapped; and that was why there was no water in the well.

❧ Charles Street Paving Gets The Go Ahead ☙

☙ **March 8, 1922**, the Board of Improvements awarded the contract for paving Charles Street to Alzina Construction Company of Springfield. The figures given were $72,000.00 and the pavement was to be of brick. The Charles Street pavement extended from the south side of Standard Addition south to Mueller's Green House on Mayo Street.

☙ **March 11, 1922,** C. B. Manback, general manager, and Thomas Williamson, company attorney, had a meeting in the Standard Oil Office on North Broad Street.

☙ Tom Paul, superintendent, and Wm. Frees, manager, passed cigars out in **March 1922,** to celebrate the new record at Mine No. 1 of 3,003 tons hoisted. Previous record was 2,876 tons.

♋ Another Coal Strike ♋

♋ On **April 1, 1922**, Illinois was hit with another coal strike, all mines in the state went idle**.**

♋ **April 12, 1922**, Thomas Williamson was in Carlinville on business for Standard Oil.

♋ **April 26, 1922,** Carlinville City Council met and adopted an ordinance allowing for the paving of Rice Street in the north part of the city. The material to be used was asphalt, at an estimated cost of $8,545.08.

♋ Charles Street Paving Gets Rehearing ♋

♋ **May 17, 1922,** the Supreme Court of Illinois granted a rehearing in the Charles Street paving matter, which had recently been decided in favor of the property owners. The rehearing indicated that the Court wished to examine some of the point's raised.

♋ **May 24, 1922,** baseball season opened in Carlinville, and the 1st game was played in Stanolind Park.

♋ Another Still in Standard Addition ♋

♋ On **June 28, 1922**, the Carlinville Democrat reported, "*At 6:00 p.m. Thursday, Officers Rasor, Staats and Poole went to the home of John Brisko, alias John Peters, of Standard Addition and made a raid where they confiscated a still, and white mule. Brisko was locked up in the county jail and Friday went before Judge Duggan charged with manufacturing intoxicating liquor in prohibition territory; he entered a plea of guilty and was fined $300 and costs. The court ordered him to jail until the fine and costs were paid. Brisko is a miner by occupation.*"

♋ **July 5, 1922,** W.E. Anderson had Friday dinner guests at his East Main Street home, C.B. Manbeck, general manager for Standard Oil, Thomas Williamson, company attorney, and C.W. Clark, local manager, as well as Judge J.B. Vaughn, Alex Crawford and William H. Anderson.

Col. Stewart speaking to the crowd of Standard Oil employees in 1920's.

♋ On **July 12, 1922,** the Board of Local Improvements let the contract for Rice Street pavement to A.M. Boatman at an estimated cost of $7,579.23. Rice Street was the 1st street on the east side of North Broad Street, south of the C & A Railroad crossing. It connects the North Broad Street pavement with Standard Addition. The pavement was to be of concrete.

♋ Coal Strike Is Ended ♋

♋ On **August 16, 1922,** the coal strike, which had been on since April 1st, ended. Operators and miners' representatives ratified a temporary wage agreement at a meeting in Cleveland, Ohio. An immediate restoration of mining was ordered in mines covering seven states.

♋ **August 23, 1922,** the coal strike was over and the blowing of the whistle at Schoper Mine was welcomed by all.

Charles Street Pavement News – Again

The **September 13, 1922, Carlinville Democrat** said, *"At the hearing of Charles Street pavement in court, it was disclosed that while estimated cost of the improvement, including overhead expenses was $80,722.68, the actual amount due the contractor was only $68,641.89. In addition to the amount due the contractor, the city included $3,626.35 for engineering and inspection, $4,569.20 for legal expenses and $2,421.66 for accruing interest on bonds. Witnesses were called by objecting property owners to find out what the $2,421.66 was for, but the city declined to furnish information. No final order was entered. The hearing on the question of benefits on the part of objectors who appealed was set for* ***September 19, 1922****."*

Schoper's Joseph Smith operating the motor on the 7-ton trolley, which gathers the coal after it has been cut [1923].

Schoper Is Open

September 13, 1922, the paper reported, *"Schoper started up last week by order of C.W. Clark, assistant manager. Miners are now at work and more will be put on as fast as room is made for them. The train service will resume as soon as arrangements are made. Berry opened a few days following settlement of the coal strike."*

Standard Oil Sued

The **Carlinville Democrat**'s **October 18, 1922,** Circuit Court News, said that a damage suit was in progress against Standard. It said that when circuit court opened that Monday, the trial of *"Arnett & Murphy vs. Standard Oil Company,"* began. *"This is a suit for damages that resulted from a sink in the ground owned by Standard Oil, which operated the coal mines near Carlinville. The lots involved, are located on the east side of North Broad Street, by the C & A Railroad. The land was purchased in* ***1918*** *by Arnett & Murphy, laid out into town lots and some of them sold. About 41 lots are alleged to have been damaged by sinking due to the mining of coal beneath. The evidence is being heard by a jury and the taking of testimony was still in progress Tuesday. The trial was still going on at press time. Two other suits of a similar nature have been filed against Standard Oil by Carlinville people."*

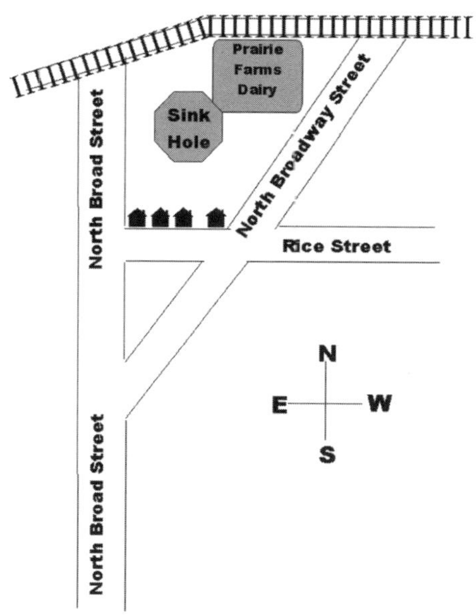

The octagonal area shaded in gray is the sinkhole site. The other area in gray is Prairie Farms Dairy in Carlinville, IL.

The sinkhole area as it looks today – The area where the sinkhole was supposedly located is outlined.

Note: It has always been told around Carlinville that the big dip in the ground by Prairie Farms, in between North Broadway Street and North Broad Street, was the result of a sinkhole. We never believed it. It really was true after all. It really is a sinkhole caused by coal mining underneath.

Oil Men Visit Carlinville

Col. Stewart

The **December 13, 1922**, Carlinville Democrat claimed that Carlinville was visited by "big" officials of Standard Oil that Friday: *"In the party of visitors were Col. W.R. Stewart, board of directors chairman ; Dr M. W. Burton, president; W.E. Warwick, 2nd vice president; B. Parks, 3rd vice president; E.G. Suebert, 4th vice president; Allen Jackson, 5th vice president; R.H. McElroy, traffic manager, E.J. Bullock; director of purchase; T.J. Thompson, general manager; C.B. Manbeck of Wood River; Thomas Williamson, their attorney of Edwardsville. In company with C.W. Clark, local manager, officials visited the mines at Berry and Schoper. E.M. Clark, of New York City, a director of Standard Oil for New Jersey, arrived later for a visit with his brother C.W. Clark. Standard is now employing a number of men at their mines and their payroll every two weeks is very large. The money that they pay to their men is spent in Carlinville and the vicinity. It is a big boost for this city and the surrounding community."*

W.E. Warwick

E.J. Bullock

R.H. McElroy

E.G. Suebert

Dr. W.M. Burton

T.J. Thompson

C.B. Manbeck

A. Jackson

Special Thanks to Thomas R. Giddens, Jackson A. Giddens & Judy (Giddens) White for use of images of the Standard Oil Company Officials. These are from their father, Paul H. Giddens' book "Standard Oil Company of Indiana – Oil Pioneer of the Midwest."

C.W. Clark

Standard Pays $110,000.00 in Wages

On **January 3, 1923,** Standard Oil paid $110,000.00 in wages to their employees for two weeks work. In January and February the total pay was estimated to be more than a half million dollars. A clip in the local paper read, *"While the coal industry is the largest, Carlinville has many other resources of importance and great value to the community. Carlinville's future is bright – conspicuously so among central Illinois cities."*

A reported accident to the tipple at the Schoper Mine on **January 17, 1923**, caused a shut down until repairs could be made.

Big Train Wreck

The Carlinville Democrat's, **February 14, 1923,** issue had an article about a wreck on the C & A Railroad. *"One of the worst wrecks that has happened in or near Carlinville, occurred last Monday night when the 11:05 – No. 6 northbound, plunged into a northbound freight south of Nilwood, IL. No. 6 made its regular stop in Carlinville and proceeded on its way. Afterwards, notice of a collision reached this city. A dense fog was evidently the cause of the wreck."*

The article further stated, *"The freight train was moving slowly, but when the engine on the passenger train hit the caboose, it caused the brakes to set on the freight and made the impact more severe. The caboose was reduced to splinters and a boxcar hauling junk iron was derailed. The engine and tender of No. 6 turned over on the left side. The express and mail cars turned over on the opposite side. All passenger cars remained on the track and aside from a severe shake up, the occupants were not hurt. Almost immediately after the crash, the boilers of the engine covered the area with steam, and the remaining parts of the caboose caught fire. The*

engineer and fireman were both discovered scalded to death by steam, and their bodies were removed. Word was sent to Carlinville, and the Schoper engine was sent to the scene, hooked onto the passenger cars and brought them into Carlinville, together with the victims. The bodies were taken to Heinz Undertaking Parlor. The mail and express clerks were uninjured. The brakeman, H.E. Wallace, was riding in the caboose of the freight and by some miraculous way, escaped being hurt except for a few scalp wounds. A tramp was riding in the coal car of No. 6 when it turned over and was caught beneath it, but managed to extricate himself. He was badly discolored with coal black. Passengers stated that not more than a moment prior to the crash, the southbound Chicago-St. Louis fast train, No. 79, passed at its usual speed on the sidetrack. It is indeed fortunate that No. 79 had passed, because the fog was so dense, that in all probability it would have crashed into the overturned engine of No. 6, which was thrown over the side of the track. A possible cause for the accident was that the engineer had withdrawn from his cabin window to let No. 79 go by and did not see the rear lights of the slow moving freight and did not realize that he was so close to it."

Rock Fall at Berry Mine

On **February 14, 1923**, Edward Glass, a miner, was injured by a fall of rock at Berry Mine.

Charles Fitzgerald, who was in charge of houses in Standard Addition, resigned **February 21, 1923.** He and Mrs. Fitzgerald moved to Chicago, where he continued work for Standard.

Right is one of many photo's that were in Sears Honor Bilt Modern Homes Catalogs during the years after Standard Addition was built.

Below is an early 1920's view of Blackburn College Campus

Note: The 20-acres that was not to be under mined is where the campus offices, buildings and dorms are located. They probably were worried about sinking under these buildings.

On **March 7, 1923**, Trustees of Blackburn College adopted a resolution to sell the underlying coal of 60-acres of the college tract – the coal underlying the other 20-acres from the college's 80 plus acre tract was not sold so that no mining can be done beneath it. The sale was to Standard Oil .

On **March 21, 1923,** C & A Railroad announced building a terminal on the old Carlinville Mine plot in the northwest part of the 4th ward, which they purchased from Standard Oil. Local reports noted, *"The work of putting up the buildings will be* *started in a short time. The roundhouse, switches and other buildings will be erected at the junctions of the main line and QC & St.L Branch. This will be an important terminal and will be a boost for Carlinville.* The Bloomington Pantagraph, that same week, contained an article, which stated "*A water tank of 100,000 gallon capacity would be located at the terminal. By installing the tank at Carlinville, it will be more convenient for crews. The nearest station at present is located at Rinaker.*"

Standard's new filling station opened on **March 21, 1923.** It was located on the corner of North Broad and First Street in Carlinville. The building was of brick and cement and the most up to date in design and construction. Earl M. Kasten was in charge of the station.

Standard Gas Station in the ***1920's***

Another Death at Berry Mine

The **May 16, 1923, Carlinville Democrat** listed another fatality at Berry: "*A fatal accident occurred at Berry on Saturday, when Charles Twitty was killed by a fall of rock. He was 33-years old. Pete Ray, who was working with Mr. Twitty, had moved just a short distance away when the crash happened.*"

Cuyler Miller, who was in charge of Standard Addition during **May 1923**, reported that every house in the Addition was full and he had a waiting list for homes.

A letter to the **Carlinville Democrat** on **June 6, 1923,** had some interesting comments: "*I remember the sensation I felt years ago when I learned that Standard Oil had bought a lot of property in and around Carlinville. The secret had been well kept. I was not much denser than most folks, but I did not know a thing about it. The few who did know kept their secret well. All of us realized that it was going to be big for Carlinville. Few, if any, realized just how big. It would take all of the space in this newspaper to tell what Standard has undertaken here, accomplished and done for Carlinville. Sometimes things can be best understood when expressed in dollars and cents. For instance, Standard had a payday recently. They have one every two weeks. It is upward of $250,000.00 a month and will soon be $300.000.00. That is about $50.00 for every man, woman and child in the city of Carlinville. Do we understand? Do we appreciate? Standard is engaged in digging coal and they have enough here to keep them at work for years. The coal is for themselves, not the market. The mines are in operation everyday. Do we understand what this means to our community? Ours is a fortunate city. Standard did not ask us to put up a lot of money to get them here, they just came. There may be slack times along business lines in other cities, but not in Carlinville. Standard Oil Company is here.*"

Still Booming

The local paper, **June 27, 1923.** "*Carlinville is surely on the boom. I drove over the city and was surprised at the new buildings being erected. Everywhere there is evidence of work. The whole county is prosperous. Carlinville is enjoying its full share. All our people are employed at good wages. It is an inspiration to view Macoupin County at this time, and it does not make any difference where you go. Carlinville is growing rapidly, is permanently prosperous. As Will Otwell said in his original manner in one of his circulars, 'These are great* *times. Great big wages. Lots of work. County of plenty. What a beautiful little city Carlinville has grown to be. Wonderful shade trees on every street in town. Beautiful avenues of hard maple and elm. Commodious hard roads over miles of our streets. Lovely homes. Solid businessmen. Solid banking, educational and commercial institutions, enlivened and helped by Standard Oil, who are paying out two or three hundred thousand dollars per month in wages. Everybody is getting some of it. It helps directly or indirectly, every citizen in town. All feel the touch of this wonderful institution that has come to Carlinville.'"*

South Broad Street looking to the south.

Old Macoupin County Jail

Left is the Macoupin County Courthouse

Truly every corner of Macoupin County did feel the touch of Standard Oil's activities in Carlinville, in some way or another.

North Broad Street in ***1920's.***

Below is East Main Street in ***1920's.***

West Main Street in ***1920's.***

The Carlinville Community High School in ***1925.***

The Carlinville Square in ***1929.*** *As you can see, Carlinville truly was and is a beautiful town.*

Another Carlinville Census

The official school census was completed on **June 27, 1923**, and the result had the population of Carlinville at 6,002. **Note:** ***Remember the last census was at 5,300 and that was in 1922.***

The Stanolinds Baseball Team defeated Witt on **July 4, 1923**, in Stanolind Park 6 to 5.

☙ Standard Addition Homes Inspected ❧

That last paragraph sounds all too familiar. I wonder if places he observed back then are some of the same ones that are like that today. I've said it a million times, it's not the people in Carlinville who care about the Sears Homes they have right in their own town, it's the people who live halfway across the world. Japanese Magazines, the BBC, History Channel, H & G, tons of media personalities and more have all done features on Standard Addition. They have interviewed and filmed individual homes as well as all of the Addition. I can't even get people to clean up front yards when they are coming. To get the city to help – what a joke! Can someone explain this to me, because I can't figure it out.

❧ On **July 10, 1923,** lawns and gardens in Standard Addition were inspected. The paper claimed, *"When Standard came to Carlinville, they did everything in 1st class style. When it came to building 156 residences for employees, they put up nice buildings with every modern convenience. The chief police officer has publicly directed property owners to destroy weeds.*

Oh how I wish this could happen today in 2005!

Standard is encouraging all employees to beautify their premises. There is considerable activity. Unfortunately, however, there are those who do not cut weeds nor care for their gardens, nor mow their lawns, nor remove rubbish. I have observed many such places in Carlinville. It's not much of a job to keep a small lot spick and span. Just a few minutes work each morning or each evening will make the yard and garden look like they were expecting company. ***AMEN!*** *The only Carlinville in the, world ought to be one of the prettiest cities in the world. Standard Addition has paved streets and the lots may not be large, but there is ample space to maintain a splendid garden and beautiful lawn. Standard sold these homes to their employees and wants to do its full share toward upkeep, building and beautifying of the city into which it has come and for which it has done so much. It is anxious that section of the city, which they are responsible for, should be neat and attractive. As an incentive to owners of homes in Standard Addition, to keep them in attractive condition, the company offered prizes for best-kept gardens and lawns. The prizes were $15.00 for best-kept garden and $10.00 for 2nd best. $15.00 for best-kept lawn and $10.00 for 2nd best. A committee made a complete inspection of all lawns and gardens in the Addition. The committee deliberated at length before final decision was made. The results are as follows: Best-kept garden – 1st place went to R.R. Thomas at 915 High Street and 2nd place went to Tony Brush at 1015 Johnson Street. Best-kept yard and lawn – 1st place went to Tony Brush at 1015 Johnson Street and 2nd place went to Thomas Carolyn at 928 Charles Street. The lawn of Sidney Duke at 934 Sumner Street got an honorable mention."*

I COULDN'T AGREE MORE! The largest collective of Sears Honor Bilt Homes in the world, ought to be given special attention – it's only 9-blocks and 152 homes!! It should be beautiful.

Street view of Sears, Roebuck and Co. houses at Carlinville, Ill.

❧ The Stanolinds won the ball game at Stanolind Park in Carlinville on **July 10, 1923**, defeating the Illinois A.C. Club of Madison by a score of 3 to 2.

❧ On **July 25, 1923, Berry** posted a notice that they would only work part time – three days a week. The apparent reason was an oversupply of gasoline.

ꟾ The **Carlinville Democrat's August 8, 1923**, issue reported, *Berry and Schoper Mines will close for the remainder of the week."*

The Jerseyville Democrat from Jerseyville, IL on **August 8, 1923** stated, *"Carlinville is making strenuous efforts to become 'City Beautiful' by exterminating weeds and rubbish. Standard offered prizes to residents in Standard Addition for best-kept gardens and lawns. Everybody, likes to see order and beauty; and weeds, vacant lots filled with weeds going to seed, piles of rubbish, do not contribute to the beauty or health of a city. Jerseyville had more weeds growing higher than ever. Since owners of weed filled premises were visited by officers and notified they must mow the weeds, the appearance of our city is 100% better."*

Berry and Schoper Mines worked steadily the entire week of **August 22, 1923.**

The **August 1923,** Stanolind Record claimed, *"John Julius and Petronilla Puleikus were married at St. Joseph's Catholic Church in Carlinville. After the ceremony, the party drove about town in decorated cars. Later, they went to the newly furnished home which the groom had prepared for his bride in Standard Addition, where a sumptuous wedding feast was served. The groom is employed at Schoper Mine."*

Note: How people raised large families in Standard Addition homes is completely baffling to me, but lots of people did and their children were none the worse for it. I guess it brought a real sense of family togetherness to be raised in a 10 children family in Standard Addition.

Injury To Berry Mine Employee

Richard Burton, of Berry Mine, was injured on **August 22, 1923,** when a car struck him and broke both legs. A lady taking her husband to work at Berry, was driving on the cinder road west of the college, when Mr. Burton stepped in front of her car and was knocked down.

On **September 12, 1923,** a daughter was born to Mr. and Mrs. John Kohn of Standard Addition. It was reported to be their 10th child.

Officers Made Raid

George Staats, David Pole, Will Homer and Marshal Rasor made a raid at 4:00 p.m. **September 26, 1923**, at the Pete Buzui residence on Rice. After searching Buzui's house, a lot of home brew, mule and whiskey was confiscated. He had a hearing before Squire Gooch and was released on $500.00 bond.

Interesting Bit of News

On **September 26, 1923,** an oil well, flowing 70,000 barrels of high gravity crude oil a day came in on West Main Street, one block west of the Carlinville Square, at Wilson Brothers Garage. Homer, the owner, invited everyone to come and see the well.

Wow, an oil well in the middle of Carlinville. Who'd have guessed?

The **September 1923,** Stanolind Record stated, *"Standard Addition looks very pretty at this time. There are a number of fine gardens and well kept lawns."* It also said, "*The company regrets the death of Louis Fuller from a fall of rock at Schoper Mine."*

Severely Injured By Fall of Coal

On **October 24, 1923,** Andrew Schroder left Carlinville for Chicago to get treatment for injuries received at Berry, when a massive amount of coal fell on him.

Special Thanks to Mary Schroder for the photo of her father.

*Andrew Schroder in his miner's outfit **1920's.***

An injury happened at Schoper on **October 31, 1923,** when Alonzo McCann was hit by a fall of rock. It injured his hip severely and he was confined to bed for a long time.

Organized Joseph A. Holmes Association

The Carlinville Democrat ran an article about the Joseph A. Holmes Association in Carlinville on **November 7, 1923.** The paper said there were more than 60 present at the meeting and a lot of visitors were in attendance as well. *"The 1st meeting of the local chapter of the Joseph A. Holmes Association was held at the First Aid training station, formerly known as Mine 1-A Office Building. The mining industry in Macoupin County has been aided by men present at the meeting. James Boston, who has been directing the First Aid and Safety First Movement for Superior Coal Co., outlined benefits derived from the work and study of preventatives and relief measures so urgently needed by every man connected with the industry. John Millhouse, state mine inspector, and Henry Knembeller, county mine inspector, brought out the need of education to dangerous conditions in the industry. They showed possible benefits by reduction of 75% of preventable accidents. Local members, Messrs. Thomas Scott, Sam Davis, Thomas Carolyn, W.E. Mason, Mike Delaney and Thomas Irvine, entertained the meeting with a repertoire of song and readings. The closing number was a real show itself as it was a wrestling match between Berry and Schoper. Berry was represented by Maurice Head, mining engineer, and Ralph Brighes, a real husky, upheld the honors for Schoper. Referee Instructor Millhouse was compelled to call time after an hour with the 1st fall undetermined and boys for each mine still pulling for their man."*

Box Social Held

An "Old and New Time" dance box supper, given by pupils and teachers at South Standard on **November 14, 1923,** was deemed a success by all present. The social was held to raise funds to purchase a musical instrument for use in the physical culture room. A large crowd attended and over 80 pies were donated for the event by patrons of Standard City and Schoper. The event yielded $100.85, which was more than enough for the instrument.

Standard Oil Mines Will Close

The **November 21, 1923,** paper reported, *"Schoper and Berry Mines will be closed. C.W. Clark, has received instructions to have the men remove their tools and clean both mines. He had orders to cease the mining of coal. Mr. Clark had no information, but a rumor was circulated about town that the mines would be sealed, but this, we understand, will not be done.*

The engineers will be retained with a small force of company men to keep the mines in good shape while they are idle. The closing is for an indefinite period."

☙ **Items of Interest Picked Up Here and There** –By J.E. McClure– About Carlinville ❧

> ***Note:*** *J.E. McClure was editor/owner of the Carlinville Democrat. This is written from his perspective.*

❧ In **November 28, 1923**, J.E. McClure's, Items of Interest Picked Up Here and There column of the Carlinville Democrat, had a lot to say about Carlinville, *"Carlinville has been growing so fast the last few years that it's hard not to realize that it was once not a city. Paved streets, sewers, new buildings, and handsome real estate, fine churches, big schools and prosperous merchants all conspire to give the county seat of Macoupin a metropolitan appearance. So accustomed are we to these things that the younger generation do not know that Carlinville lacked a lot of being a city some years ago. Some say that it is interesting and helpful to the community to recall what it used to be. The world does move and new supplants the old, but this all marks an epoch in Carlinville's interesting history."*

☙ Standard Oil Company Will Not Sell ❧

❧ On **December 19, 1923**, it was reported, *"Since the closing of the coal mines in this city recently, there has been much curiosity among our citizens as to why it was done and not a little anxiety, lest it be permanent. In compliance with a wide spread desire to know what the status of the case is, Mayor Hemphill went to Chicago and called upon the chairmen and board of directors of Standard Oil. Col. R.W. Stewart, and members of his board, received him cordially and listened attentively to Mayor Hemphill's presentation of Carlinville's side of the situation. Col. Stewart, very frankly, told him that there is a surplus of crude oil, and likewise, coal, which Standard Oil uses. 'Crude oil is so cheap that it is being used by many railroad companies for fuel and this in consequence reduces the demand for coal. There is, accordingly, a surplus of coal on hand, said Col. Stewart, Standard Oil, as a business proposition, could not afford to mine coal, when they could buy it at much less than cost of production in open market." The mayor was assured that Standard Oil had no intention of disposing of their interests in and near Carlinville, and secondly, that they will continue to keep the mines in such good condition, that they could be ready for operation within 10 days."*

❧ On **February 6, 1924,** the Carlinville Democrat said, *"The special committee on roads and bridges, composed of Supervisors Dewitt, McCollom, Behrens and Cooper, have hired an engineer to survey the road to Palmyra, IL. Forrest Nifong, engineer for Standard Oil, was 1st employed, but he was called to Wood River by the company and had to give up the position. Schwab & Company of Alton, IL was hired to take Mr. Nifong's place and are now at work making the survey. The road will be hard surfaced and will comply with all provisions of the appropriation of $51,000.00 which is available for that purpose."*

☙ Schoper Will Open ❧

❧ A news item on the front page of the **April 23, 1924**, Carlinville Democrat stated, *"Schoper Mine will open May 1st. This was announced by officials of the company in this city following a meeting held at Schoper. The meeting was held to go over matters concerning conditions in the* *mine and was settled, with some details that have to be approved by higher officers of the union. The mines closed in mid* ***November 1923****, and about 1,000 men were out of work. It was a severe blow to Carlinville and news to resume work was received with much rejoicing by businessmen. Following the meeting, C.W. Clark, manager, stated that 10 men would go to work to do necessary preparatory work for the start of regular mining on May 1st. At the beginning, it is hoped to mine and ship 1,500 to 2,000 tons of coal daily. As things get into shape, they will increase the number of men and it is thought that within 60 days, 600 men will be at work and output will be 4,000 tons daily. This means about 100 railroad cars per day. The train from Carlinville to Schoper will commence running May 1st. Berry Mine will not be opened at present. It is understood that cost of operating that mine is very high and it has been expressed that there is a possibility that the mine will remain closed."*

ଓ Whistle Anticipation in Carlinville ഇ

ഇ People in Carlinville were anxiously awaiting the sound of the whistle blowing at Schoper. The **April 24, 1924,** paper said, *"It will announce to the world that Standard Oil's coal mine is in operation once again. It will have far reaching effects on the community."*

ഇ **April 30, 1924,** *"Tomorrow will be a significant day in Carlinville and Macoupin County. Schoper will open after being closed for 5 months. Our people are happy that the industry will again be in operation. The clouds have rolled by and the sun is shining on Carlinville."*

Miners at Schoper waiting to go down to work.

ଓ Schoper Mine Is Open ഇ

ഇ The Carlinville Democrat reported on **May 7, 1924**, *"Schoper opened May 1st. There was great activity at the mine when men began to go down at 6:30. A number of men who had worked at Schoper and had left Carlinville for employment elsewhere, are returning and going back to work at Schoper. At present, the total is 330 men at work. Coal is being hoisted now and the output will be increased day to day. The C & A special train left the station at 6:00 a.m. and it seemed like old times. It is indeed a splendid thing for this city that the mine has reopened and already there is a much brighter outlook in business circles in the county seat."*

ഇ On **June 7, 1924,** Schoper set a new record of hoisting 4,022 tons of coal.

ഇ Atty Williamson was in Carlinville **June 11, 1924,** conferring with officers of Standard Oil.

ଓ More Injuries At Schoper Mine ഇ

ഇ **August 20, 1924**, Albert Hoppe and Thomas O'Neil were injured at Schoper. Mr. Hoppe's back was injured and Mr. O'Neil's hand was smashed when a loaded coal car hit it.

☙ Returns To Work At Schoper Mine ❧

❧ After being off work 10 days with a broken bone in his toe resulting from an accident in Schoper, Thomas Sweeney returned to work on **August 27, 1924.**

❧ The **August 1924,** Stanolind Record reported, *"Schoper has again resumed operations with a larger tonnage than before the shutdown."*

❧ The **September 1924,** Stanolind Record claimed that the latest record for hoisted coal at Schoper was 4,030 tons in 8 hours. It also said, *"Mr. and Mrs. Harold Reineke, Mr. and Mrs. J.L. Gemp, Charles VanMeter and family and George Loehnberg and family had all moved from Carlinville to their summer villas at Schoper Lake."*

☙ Schoper Mine Idle Again ❧

❧ On **October 1, 1924**, the local paper reported that Schoper had been idle from Wednesday to Saturday the previous week, but would start work again on Monday.

☙ Suicide in Standard Addition ❧

❧ At 4:00 a.m., **October 22, 1924,** Night Officer George Staats was called to George Kassa's residence on the corner of Rice and Charles Streets in Standard Addition. Upon arrival, he found Mrs. Kassa on the sidewalk a half block from the home. She told Officer Staats that her husband had fired 2 shots and she did not know what happened. Mr. Staats went into the house and in an upstairs bedroom he found Mr. Kassa on the floor with a bullet hole just above his right ear. Mr. Kassa was dead. Mrs. Kassa said that her husband had left home at 3:00 a.m. When she asked him where he'd been, he refused to tell her. She said he was under the influence of liquor. After he'd been home a time, she tried to persuade him to go to bed. He refused at first, but finally lay down on the bed. She got up to go downstairs, but before reaching the bottom step she heard 2 shots. She didn't go back to the house until Officer Staats arrived. At the head of the stairway, was a bullet hole through the bedroom wall pointing down. It is believed that Kassa fired at his wife as she was going downstairs, missed and then shot himself. The gun was found on the floor a foot from Mr. Kassa's right hand. Neighbors said the Kassa's had been arguing for weeks. Mrs. Kassa said he frequently threatened to kill her. He was employed at Schoper.

Note: Rumors are that Mr. Kassa haunts this house. I wasn't able to determine which home this was. It could be any of the four homes that are on the corner of Rice and Charles Street.

❧ The **November 1924,** Stanolind Record said, *"The new record for No. 2 Mine is 4,210 tons hoisted in 8 hours."* It continued, *"A.C. Wolf and family have taken the last house at Schoper."* It stated about Standard City, *"School opened **September 2**, with Misses Elizabeth Behrendt and Nora Griffin as teachers; both reside in Standard City."*

❧ The Carlinville Democrat's, **November 19, 1924** edition stated, *"Standard worked their mine 3 days last week and 3 days this week."*

☙ Schoper Closes Again ❧

❧ The Carlinville Democrat reported on **December 10, 1924,** *"Schoper Mine closed indefinitely.*

It's regretted that the mine is closed and is hoped that it will not be for any great length of time."

ଓ Nilwood Mine Closed ଛ

ଛ Mrs. B.T. Cullop of Nilwood, IL was in Carlinville **January 14, 1925,** and said, *"The coal mine at Nilwood has been closed for a time. There are no prospects for it to be reopened. Some miners who worked there and at Schoper are working at Solomon Mine near Virden."*

ଛ The **January 1925,** Stanolind Record reported, *"There have been several new building contracts awarded lately; among these are John Kroeger's new residence and John Swanson's corn crib. Both buildings are expected to add to the beauty of Schoper Lake."*

ଓ Train Hits Ford Truck ଛ

ଛ Wednesday **February 11, 1925,** at 10:00 a.m., a train hit George Underkoffler's truck . The truck was demolished. Mr. Underkoffler was seriously injured. The accident occurred by the Schoper Power House, at the top of a steep grade. He was employed by Standard Oil.

ଓ Schoper To Reopen Again ଛ

ଛ Men were busy cleaning the shaft at Schoper **March 11, 1925,** getting it ready for the diggers. The paper reported that, *"Schoper reopened after being closed since December. This is the best news received in Carlinville for months. At one time, more than 600 men were employed there. There was a large payroll every 2 weeks, which was a great benefit to the merchants and citizens of Carlinville and vicinity. Owing to the fact that screenings could be bought for less than the company could mine at Schoper, it was necessary to close the mine. The officials said that it was losing them a large sum of money. C.W. Clark, general manager of the mines here, said in regard to the opening, 'Schoper Mine opened Monday and men are at work getting the mine in shape. As fast as conditions warrant, miners will be put to work and we hope to be running at full capacity by April 1st.' Standard has done much for Carlinville. It is great news to people of this city and vicinity to know that our big mine has started and as far as anyone can tell, will run steadily all summer and it is hoped, up to the middle of winter."*

ଛ **March 11, 1925**, Schoper's whistle blew, signaling the mine opening.

ଛ C.W. Clark, superintendent, was confined to his home on **March 11, 1925,** with an extreme attack of the flu.

ଛ On **March 18, 1925,** 450 men went to work at Schoper. The Carlinville - Schoper train left Carlinville at 5:45 a.m. to take the miners to and from work at Schoper. C.W. Clark recovered from his bout with the flu and returned to work.

ଛ On **May 6, 1925,** Standard City School held a successful indoor carnival, with proceeds bringing in around $175.00. There was a very large crowd in attendance.

Death At Schoper

35-year-old Glenn Savage was killed on **June 17, 1925,** in Schoper when he came in contact with a live wire and was electrocuted.

The **June 1925,** Stanolind Record reported, *"Rick Kelly, engineer at Schoper, was elected to represent the 5th ward at City Council."* It also stated, *"E.R. Baker, chief engineer at Schoper Power Plant, said it took a while, but he caught a big huge catfish at Schoper Lake. When asked by the reporter how long he was, Mr. Baker said 'about 2 hours.'"*

Houses In Standard Addition To Be Painted

Alex Gaspard was given a contract to paint houses in Standard Addition on **July 15, 1924**. It was understood, that Standard wanted all of the houses painted.

Houses in Standard Addition to Carlinville in 1920.

Interesting News

In a Carlinville Democrat interview on **July 15, 1925**, John F. Kiefer of Carlinville said, *"Last week someone was talking about the old brass cannon, located in the park on the square. This cannon was used years ago to usher in the 4th of July and other occasions. I remember years ago that there was a celebration being held at the courthouse. The cannon was hauled over to the courthouse to help celebrate. At the time the celebration was going on, they were plastering the Presbyterian Church. The field piece was fired in grand style, but with bad results for the church. The new plastering, which was not dry, all came down and the blast of the gun also cracked windowpanes in the courthouse."*

Note: Even though this item didn't relate to the Standard Addition Saga, I found it doing research for this book and thought it interesting. We've always been told that the cannon was never and couldn't ever be fired – guess who ever told that was wrong.

The infamous cannon – still sits on the Macoupin County Courthouse lawn today.

North Side Miners Store

In the Carlinville Democrat **July 22, 1925**, attention was called to the advertisement of North Side Miners Supply Company of Carlinville. The paper said, *"They are*

*Building to left in the photo is the North Side Miners Supply Store; this was built when Standard Addition Houses were in **1918 – 1919.***

having a sale at their store on Rice and High Streets and it is a big success. Patrons who take advantage of this sale have a chance to get a $150.00 radio set, and 3 burner Perfection Oil stove free. The sale ends August 1st."

☙ Standard Oil To Shut Down Mines ❧

❧ Standard announced on **July 29, 1925,** that they would shut down. The paper reported it as *"A severe blow to Carlinville, but fortunately, Carlinville was never dependant upon coal mining. Carlinville is the center of a wonderful agricultural region. Crops are good, and if proper effort is put forth, additional trade will be secured for our city which will be sufficient to more than make up what will be lost by the closing of the mines."*

Sad Miners.

After just a few short years, Carlinville's big boom would go BOOM! The mines were closing permanently, although Carlinville didn't know it at the time. The growth that our town had been experiencing would stop.

Never again would Carlinville see anything like the historical events that had taken place in the 8-years that Standard Oil Company had made their presence here. Was Carlinville better off because they had come or would it have been better if they hadn't?

When Standard Oil announced that a shut down of the mine was going to take place, the local paper, as well as the citizens of Carlinville said, *"Fortunately Carlinville was never dependent on the coal mining industry."*

Were they really not dependent on the coal mining industry and could Carlinville now survive without it?

Would everyone see just how much the Standard Oil Company had meant to Carlinville? Maybe they would, but maybe, it would also be "too little - too late."

☙❧

The look on the miners face in the photo to the right, says it all –
What do I do now?

Chapter VIII

"The Big Boom Goes Boom"

The Big Boom Goes BOOM

☙ Standard Oil Closes Mines ❧

❧ On **July 29, 1925**, the Carlinville Democrat reported that Standard would close its mines in the Carlinville area, *"Rumor has it that Standard will close Schoper Mine on August 1st. Officials have made no statement, yet it seems understood that such action will be taken. No coal is sold on the market, the whole output is converted for use in the refineries. When screenings can be purchased for considerably less than can be mined, Standard ceased coal operations. It is likely that the mine will not reopen prior to* ***April 1, 1926****."*

☙ Schoper Train Has Wreck ❧

❧ The Schoper train, which carries men to work at the mine had a wreck on **July 29, 1925**. It left Carlinville everyday at 5:45 a.m. The wreck occurred near Schoper Power House. To avoid hitting coal cars, the emergency brakes were put on suddenly. Some of the men were bruised and shaken up and one man received a broken collarbone. The wreck occurred at 6:45 a.m.

☙ Interesting Bit of News ❧

Early 1900's View of the Macoupin County Jail

❧ The Carlinville Democrat of **August 5, 1925** claimed, *"Mob Violence Was Feared in Carlinville. About 50 men assembled near Carlinville. Sheriff John Russell was fearful the a mob might take custody of Paul Blackburn and Charles Lewis, both held in jail, charged with the murder of William Barnes of Nilwood, IL. He took them to jail in another county for safety and recently returned them to Carlinville. Last Friday night he learned that men had assembled north of Carlinville near Berry Mine and were intending to storm the jail and seize Blackburn And Lewis. He swore in a number of deputies and put them on guard at the jail.*

Sheriff Russell and City Marshal Clarence Rasor went and talked to the men, most lived in the vicinity of Nilwood. The sheriff asked them to disperse. He returned to the city and thought the men were going to proceed with their intentions of seizing the prisoners. The sheriff appealed to Mayor Hemphill who talked to them and got them to disperse. The affair caused considerable excitement. Members of the party who assembled were all friends of William Barnes and law-abiding people. They want justice to be meted out promptly to the two men. They were convinced to allow the law to take its course. Macoupin County Jail is a very substantial one and it would be an impossibility to break through the iron locked and barred doors of that institution except after hours of laborious effort. Unless they could secure the keys, it would be a futile and unsuccessful endeavor." The two men were charged with the murder of Deputy Sheriff William Barnes at Nilwood on **July 20th.** The men gave their names as Paul Blackburn and Charles Lewis and said they were from Chicago. This was only part true as they were from Chicago, but were wanted in East St.Louis. Blackburn gave his right name, but the other was named Charles Lewis Orsborne. When arrested, Blackburn had a bullet wound in his neck and shoulder. Blackburn was 19 and Lewis 22. The hunt for the men continued from the time Barnes was killed up to the time of their arrest. It was the 1st time an airplane was used for a manhunt in Macoupin County. They were also wanted for the robbery of Albert Campbell, D.B. Hughes and wife, and Ray England.

Note: Even thought this item had nothing to do with the Standard Addition Saga, I had always heard about the "big lynching mob" growing up and when I came across it while doing research for this book, I couldn't resist putting it in. A "lynching mob" in Carlinville – WOW! – It really was true.

Gillespie Coal Mine Closes

On **August 5, 1925** Superior Coal Mine near Gillespie, IL closed for the same reason that Standard closed Schoper. *"Coal can be bought much cheaper than it can be mined."*

Gillespie Coal Mine in 1924.

The **August 1925,** <u>Stanolind Record</u> claimed that, *"When John Perino returned from work recently, he found his home in South Standard nothing but charred embers. The fire started from a coal oil stove."* It also said, *"The flume at the power house was drained recently. As the water got low, the people that were watching jumped in swimming and fishing. Most came out with wiggling victims in their pockets and inside their shirts or anything else they could use to hold the fish."*

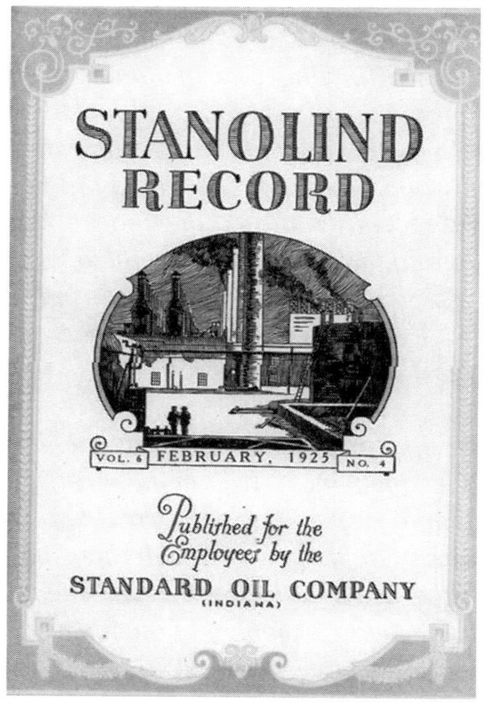

Note: The above article was the last item that appeared in the <u>Stanolind Record</u> about Carlinville, Standard Addition, Schoper and Standard City. After the mines closed, Standard Oil really did want to forget about us.

Standard Will Not Lease Berry Mine

On **September 30, 1925**, Carlinville miners who had worked in Berry Mine made a proposition to the company to lease the mine. C.W. Clark, general manager of Standard for Carlinville, went to Chicago to see the head officials, but they declined to lease the mine.

The Standard City Band came to Carlinville on **September 30, 1925**, and gave a big concert in the park from 7 to 10 p.m. It was their 1st visit to the city as a musical organization and they were said to have, *"Completely won the hearts of the large crowd."*

*Above is a **1925** postcard view of the Carlinville Square. To the right is a close up of the gazebo area where the bands performed.*

C.W. Clark, manager of Standard's interests in Carlinville, spent all day, **November 4, 1925**, in Wood River on business at offices of Standard Oil.

All offices of Standard Oil Company moved from Schoper to their office building on North Broad Street on **November 4, 1925.**

An entertainment and supper was held at the Standard City School on **December 18, 1925.** They had a vaudeville show, plenty of music and lots of good things to eat.

General Manager Manbeck was in Carlinville on **January 20, 1926,** inspecting Standard Oil Mines. The local paper stated, *"No intimation was given out as to when the mines might open. Mr. Manbeck and his assistant manager, C.W. Clark, had quite a conference, and the public generally hope that the mines may open in the near future."*

On **March 10, 1926,** Standard ran an ad advertising their Polarine Motor Oil available at their filling station in town. Standard called their oil *"the perfect motor oil - the language of squeaks."*

Miner and Family Leave Standard Addition

Mr. and Mrs. Jenkins, who had resided in Standard Addition since the building of the houses, moved on **March 10, 1926** to Taylorville, IL. Mr. Jenkins was a miner at Schoper.

Standard Oil Officers Met

On **May 12, 1926**, C.B. Manbeck, superintendent for Standard, spent the day in Carlinville. He was in conference with C.W. Clark, resident manager, and Thomas Williamson, company attorney. Assessments and taxes were discussed.

☙ Survey of Carlinville ❧

❧ The **May 19, 1926, Carlinville Democrat** ran a huge article about Carlinville being the only Carlinville in the world: *"The Carlinville Chamber of Commerce prepared a survey of Carlinville. A committee was appointed to compile a brief statement, of information contained in the survey. The following is the result: Carlinville is the County Seat of Macoupin. It is located on the main line of the Chicago & Alton Railroad, 224-miles southwest of Chicago, IL, 39-miles southwest of Springfield, IL and 59-miles northeast of St. Louis, MO. It is the center of one of the largest counties in Illinois. Carlinville's natural resources are abundant and easily accessible. It is surrounded by a huge coal field, has limestones in large quantities near at hand, and clay and shale. It has a splendid water system. It has an excellent sewer system. It has an electric power plant and gas factory, which furnish power, heat and fuel. Its three banks have a total resource of nearly three and one half million dollars. Its population is 5,300, but its advantages and attractive features are usually found only in cities twice that size. It has a hospital. It has a Rotary Club, Women's Club, a Chamber of Commerce of nearly 200 members and many social and fraternal organizations. Carlinville is a fine place in which to live. Its streets are beautifully shaded and well kept. Seven miles of them are paved. It is a city of good order and of splendid ideals. It has long been famed for its educational facilities. It is the home of Blackburn College. Carlinville's Community High School is one of the finest in Illinois.*

*Tree lined Carlinville Street in early **1920's.***

There are nine protestant and two Catholic churches. Carlinville is both a good place to live and to do business. It has a number of industries in successful operation. Among them are a tile factory, a brick factory, the Starr Brother's Creamery, the coal mines of Standard Oil where necessary fuel for three refineries is produced, a small coal mine, which is a local enterprise, which supplies the local community at a minimum of cost, and three successfully operated dairies. But other industries are needed. Soil conditions surrounding Carlinville are such as would produce all of the fruits and vegetables needed for a large cannery – an institution for which there is a special demand. Good sites for industrial purposes are obtainable, located in close proximity to railroads.

It is the only Carlinville in the world."

Right is old St. Joseph's Catholic Church and above is old St. Mary's Catholic Church.

High School, North School, Blackburn College and the Macoupin County Courthouse in the ***1920's.***

Charged With Stealing

On **July 21, 1926,** City Marshal Rasor arrested six boys, 5 to 11 years of age. All were questioned in regard to a robbery at the soft drink parlor owned by Frank Silvarstrucci and Jake Pagani in Standard Addition. The boys were charged with taking $85.00 from the cash drawer.

Photo to right is of the old Soda Parlor in Standard Addition; it is on the right.

Quite a baseball game was played at Stanolin Park **July 21, 1926** between the Bloomer Girls and the Carlinville Team. The Bloomer Girls won 9 to 7.

Big Explosion In Standard Addition

On **August 1, 1926,** the Carlinville Democrat said the city was shaken by an explosion: *"A two story building in Standard Addition was destroyed by an explosion. It was followed by a fire and authorities were closely investigating as much insurance was carried on the property. At 2:30 a.m. citizens were awakened by an explosion, which shook houses and rattled dishes. Except for the loud report that accompanied it, it was like a slight earthquake. A fire alarm was turned in and it was found that the wholesale grocery house in Standard Addition was in flames. The building was blown to pieces and on fire. The destruction was complete. The building used to be the soft drink parlor. Recently, it has been occupied as a wholesale grocery house, under the name of Macoupin Wholesale Grocery Company. Since June 1st, the proprietors have been Leonard Ciaccio and Joe Scaparia of Springfield, IL. They carried a large stock of groceries. Both floors and the basement were filled. That the explosion was a severe one is evident to any who felt the effects of it, or looked upon the destruction. Considerable damage was done to other* *property in the immediate neighborhood. Windows were broke, building material was blown through the weatherboarding of buildings and residents in the neighborhood were greatly shocked. Several were thrown from their beds to the floor. Following the explosion, the wreck was completely enveloped in flames. Firemen spent the night trying to subdue the fire. Included in the destruction were an REO Truck and a Lincoln Touring Car. There was considerable insurance on the 2-story building and stock. As far as we have been able to learn, the following is the insurance amount carried: In the Ed O'Neil agency - $3,000.00 on building, $500.00 on garage, $15,000.00 on stock. In the Westermeier agency - $10,000.00 on stock. In the Dick Dunn agency - $2,000.00 on building, $1,000.00 on stock. Everyone agrees that there was an explosion. There is nothing conclusive as to what caused it. That an investigation will be made is certain. The insurance companies will have their representatives here and the state fire marshal's office will no doubt conduct a searching inquiry."*

The **August 11, 1926**, **Carlinville Democrat** said, *"The explosion and fire that destroyed the Macoupin County Grocery Company in Standard Addition last week, is being investigated by both the insurance companies and the state fire marshal's office. James Wyse, deputy state fire marshal, is in Carlinville and supervising removal of debris. He has not said if the real cause of the explosion has been ascertained. All are agreed that the loss on the building was complete."*

Carlinville Fire Department was called to Standard City on, **September 15, 1926**, at 10:00 a.m. Coal stored in the Schoper Power House had caught on fire. The blaze was extinguished without difficulty.

Booze Plants Taken at Standard City

The **Carlinville Democrat** of **January 19, 1927,** claimed *"Sheriff Stubblefield and officers made a big haul at Standard City and estimated that 100,000 gallons of mash were destroyed and 3 men arrested. Sheriff Stubblefield and his deputies motored to Standard City and made a raid on 2 white mule factories. They captured 2 of the largest stills ever uncovered in Macoupin County. The 1st place raided was a 4-room stucco house. At this house, the officers knocked on the front door and there was no response. Further investigation at the rear disclosed a building, which contained an enormous "double deck" still. The officers discovered 3 large rooms under the house that were filled with mash. On returning to the rooms above, they found that each room had a door in the floor, about 3-feet square, which had been sawed out to make the opening. From this opening, the mash was stirred with wooden paddles. The size of the basement, which was converted into 3 vats, was 32 by 30 by 8-feet.*

The next place raided was 4 or 5 blocks north and 1 block east. From a distance, it looked like a small elevator. The basement of this building was 15 by 30 by 8-feet and made of concrete. It was filled with mash. In the building, there was a 3 column still set up to make alcohol. When officers drove up, 3 men ran out of the building and were immediately arrested. They offered no resistance and no weapons were found in the building or on the men. The men arrested were Tony Calcaro, Murzo Brunello and John Biagaido. When the premises had been thoroughly inspected and the men taken into custody, Sheriff Stubblefield returned to Carlinville and the 3 prisoners were placed in jail. The other deputies remained at Standard City to guard the property. When Sheriff Stubblefield returned to Standard City he dynamited the concrete *basement. Kerosene was poured into the pipes and on what mash remained in the bottom of the vats. Wednesday afternoon, the 3 men were released on $4,000.00 bond each.*

Sheriff Stubblefield loaded the stills on two large trucks and took them to the basement of the courthouse. The plant at the 1st house was used for making mule; the other was used for making alcohol. At one of the places, water for the mash was taken from wells. At the other, water was supplied from the Schoper Reservoir through a rubber hose about 2 inches in diameter. About 200-feet of the hose was brought in by the sheriff. Apparently, the still in the 1st house had been running for a long time. The other had only been in business a few weeks. A small quantity of liquor was also found. Officers formed the impression that the liquor had been transported to other cities and very little of it was sold in Standard City. It is believed that the 3 men arrested had no ownership in the stills."

☙ Winter Blast Hits Carlinville ❧

❧ Wednesday, **January 19, 1927,** a winter blast hit Carlinville. The thermometer fell to 16 degrees below zero and 6 to 10-inches of snow covered the county.

☙ Another Mine Subsidence Suit ❧

❧ On **January 19, 1927,** Attys James H. Murphy and Victor H. Hemphill went to Chicago for a hearing before the United States Court of Appeals about the case of Watts vs. Standard Oil. Murphy and Hemphill represented the Watts family. The suit was heard months before in the United States District Court in Springfield, IL where a judgment was rendered of $5,000.00. This represented damages to the Watts farm north of Carlinville, due to subsidence. Standard Oil appealed the judgment.

❧ The 3 men arrested in the raid on Standard City **January 19,** were charged on **January 26, 1927** with manufacturing and possession of intoxicating liquor. A petition was filed by their attorney to quash search warrants, suppress evidence and restore property, which was seized. It was filed before Judge Issacs of county court and was to be heard at a later date.

*Ice storm that hit Carlinville in the late **1920's**. The photo is of the city square.*

☙ Winter Hits Again ❧

❧ Old Man Winter visited Carlinville again on **January 26, 1927**, when the city and county was hit hard by sleet and freezing rain. The snow that had fallen a week earlier, was covered with a thick coating that was as slick as glass. Walking was difficult and dangerous, driving a car was impossible. The traction line was not able to run as ice coated the trolley and wire. An icy glare covered trees and wires around town.

It was so thick and temperatures so cold, that it wasn't expected to begin to thaw for a week.

Carlinville Trolley & Traction - **1910.**

ꟼ Coal Operators and Representatives of United Mine Workers held a meeting **February 16, 1927,** In Florida, which brought rumors to Carlinville that Standard would open their mines near Carlinville. No verification was obtained from Standard.

ꟼ On **April 13, 1927**, Rollin Smith went to Chicago and resumed work as traveling auditor for Standard Oil. Mr. Smith moved his family from Carlinville to Chicago later.

ꟼ C.W. Clark, manager of Standard's properties in Carlinville, left **May 18, 1927,** for Cincinnati, OH to attend the American Mining Congress.

Sue For Fire Losses

ꟼ On **August 3, 1927,** Macoupin County Grocery Company of Carlinville, owned by Leonard Giaccio and Joe Scaperia, filed a lawsuit in Sangamon County, against 5 insurance companies insuring their building and properties at the time of the explosion. The suit compelled the insurance companies to pay the losses.

ꟼ A number of windows in vacant houses in Standard Addition were broken out on **August 3, 1927**. Damage was estimated at $180.00. Seven boys were taken into custody and questioned. They admitted to the damage and were taken before State's Attorney Harlan, who gave them a stern talk on the seriousness of their offense and paroled them to Marshal Rasor where they were to report every 30 days.

Big Hole Opens Up On Charles Street

Note: This crossing was on Nicholas Street where Charles Street crosses it. The tracks that ran along Nicholas Street have been removed.

ꟼ City authorities discovered on **August 3, 1927**, that there had been a washout under the railroad tracks where the Charles Street pavement crosses. The heavy rains of the previous months had run the water past the inlet of the tile and had undermined the track. According to the paper, *"A space 20 by 8 by 10-feet deep was washed out on the QC & St.L branch of the C & A Railroad. Those who saw it, wondered how the locomotive crossed without causing a cave in. A carload of cinders was used to fill the hole."*

Dorm Fire at Blackburn College **1927.**

Interesting Bit of News

ꟼ The **August 24, 1927,** issue of the **Carlinville Democrat** reported, *"A disastrous fire at Blackburn College. The building known as the boys' dormitory and administration building,* *was destroyed by fire. Two students, who were rooming on the 1st floor, discovered the blaze around 1:00 a.m. They gave the alarm and the fire department responded promptly. The firemen worked heroically, but could not extinguish it. A number of citizens went to the scene and helped save a large number of books and records located on the 1st floor. It is thought that crossed wires started the fire. The building is a total loss. Professor Challacombe, instructor and bursar for Blackburn, said $55,000.00 insurance was carried on the building. He also said, the Science Hall north of the structure will be used for temporary classrooms and also portable school buildings in Standard Addition if necessary."*

ജ By **September 4, 1927**, the new Blackburn building was under construction, the paper said, *"A lot of work is being done at Blackburn College. Considerable excavation has been done already for the new dormitory. It is going to be of the same design and size as the girls' dormitory. The new building will be located near the west side of the campus. It will be built with student labor, just as Stoddard Hall was built. Rapid progress is being made in transporting the Standard Addition portable school buildings to the Campus. President Hudson is confident that these buildings will be in place and ready for use on opening day of college."*

Above are Standard Addition portable school buildings after being moved to the college campus. Left are Standard Addition portable school buildings just before being moved, if you look closely you can see Standard Addition houses in the background.

ജ Blackburn College opened on Monday, **September 19, 1927**. One of the wooden school buildings from Standard Addition, which was loaned to Blackburn, was used as an office building, while the other was used as a classroom. A couple of Pullman cars on loan from the C & A Railroad were divided into classrooms.

Blackburn classes held in a Pullman car in ***1927.***

ജ

Left are Pullman cars being moved to the Blackburn Campus - ***1927.***

Blackburn students beside a Pullman car used for classes.

By now, Standard Oil mines near Carlinville had been closed for a while. The citizens of both Carlinville and the surrounding communities were wondering if they ever would reopen.

Was Standard Oil Company Gone?

The houses in Standard Addition were almost all vacant and the Addition looked like a neglected old ghost town. It was falling into such a bad state that people in Carlinville were calling it "Sub-Standard Addition." For the 1st time, reality that Standard might never come back to Carlinville, was hitting the city hard. People were faced with the fear that one of their largest industries was gone and beginning to realize just what Standard Oil Company meant to the area. The Carlinville community would not let this much-needed industry go without putting up a fight.

On **January 25, 1928,** the Carlinville Democrat ran articles calling attention to citizens of Carlinville and the area, about a coming meeting at the courthouse to see what could be done about Standard Oil Company and the situation regarding them in Carlinville.

*Macoupin County Courthouse in the **1920's.***

Could Carlinville get Standard Oil to come back?

Could they strike up some kind of bargain with them to get the mines open?

Could they even talk them into selling the mines and let someone else run them?

Surely there had to be something they could do.

Times were desperate and required desperate measures.

Could anything be done?

Chapter IX

"Are You Coming Back Standard Oil Company?"

Are You Coming Back Standard Oil Company?

ꟾ *Are you coming back Standard Oil Company?* This was the question many people in Carlinville had on their minds in **1928**, when the Standard Oil Mines near Carlinville had been closed nearly two years. Carlinville citizens decided to put up a fight to bring the much-needed industry back to the area, but what could they do?

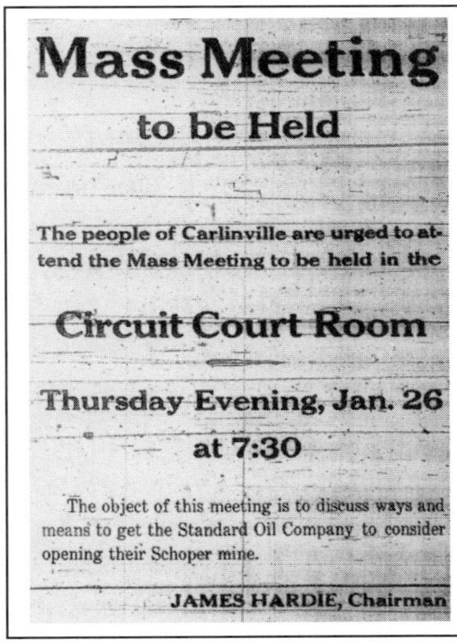

Mass Meeting

to be Held

The people of Carlinville are urged to attend the Mass Meeting to be held in the

Circuit Court Room

Thursday Evening, Jan. 26

at 7:30

The object of this meeting is to discuss ways and means to get the Standard Oil Company to consider opening their Schoper mine.

JAMES HARDIE, Chairman

Appeal To Standard To Open Mine

A meeting was held **January 26, 1928** in the main courtroom of the courthouse to discuss the situation with Standard Oil and the ways and means that Carlinville could go about to secure the opening of Schoper Mine. Several hundred citizens were present. James Hardie presided. The newspaper reported, *"The purpose of the meeting was to let Standard Oil know that Carlinville people are anxious that mining operations resume and are eager to cooperate in every possible. There were a number of speakers. Mr. Hardie 1st presented Mayor Victor Hemphill, who pointed out the things that had been done by Carlinville, during the time that the company was operating their mines here. Sewers were constructed, water mains laid, electric lights extended, pavements built, houses erected and the right hand of welcome was extended to thousands who came to Carlinville as part of the great coal activity. Following the Mayor, Frank A. Muench, a businessman of Carlinville and Chamber of Commerce President, spoke and promised his cooperation in any way that would bring about a renewal of coal mining activities here. E.A. Ibbetson, J.E. McClure, C.J. Lumpkin and E.H. Yarke, district freight and passenger agent of the C & A Railroad, all spoke briefly and pledged cooperation. Representative Whiteley read a set of resolutions relating to the matter, which had been discussed and moved for adoption. The motion was promptly seconded and prevailed without a descending vote. Upon motion, the meeting was adjourned. The resolutions adopted at the meeting are as follows:*

When Standard Oil of Indiana established coal mining in Carlinville, this community experienced a considerable boom and our people had a vision of rapid and permanent growth.

We were pleased that Standard Oil had selected this town as the location of one of its principle activities. We were intensely proud of the mines, the power plant, the office building and the homes it built for its employees.

We endeavored to meet requirements and public and private improvements were constructed. New stores opened, larger stocks of merchandise were acquired and people strived to do everything to provide our new population with all conveniences and advantages to which they were entitled. Many people were attracted to Carlinville and obtained employment. Many were so impressed that they invested in homes and moved into them with hope that they were permanently located. Then came the closing of the mines, resulting in loss of employment to a large number of men. Many families had to leave their homes and go elsewhere in search of employment. Then the vital importance of the Standard Oil mining industry to this community became magnified and intensified by its suspension. Our people did not give up. They have carried on and made valiant efforts to make the best of a hard situation. We would like to express to Standard Oil, our hopes that the industry will again become our chief industry and our desire to cooperate with the company and its employees that its mines may be reopened and kept open."

"BE IT RESOLVED BY PEOPLE OF CARLINVILLE assembled in this mass meeting:

That we now give recognition to the value of Standard Oil Company as an enormous factor in the prosperity of this community; and that we do here and now pledge to Standard Oil Company, our continued and loyal cooperation and support in all endeavors to reopen the mines here and to keep them open; and that we do appoint James Hardie, Edward Lott, Earnest Ibbetson, W.M. Hudson and Victor Hemphill, citizens of Carlinville, as a committee to carry these resolutions to the Board of Directors of Standard Oil Company of Indiana and to convey to that board the good will of the people of this city, and to request that board to resume its mining operations here just as quickly as possible, and hope that through closer association with the resumption of activities here, our people and officials of Standard Oil Company of Indiana may become better and more cordial friends."

ℵ On **February 8, 1928**, Carlinville City Council met and a petition was filed asking that a light be placed at North Broad and Rice Streets. It was referred to the Public Utility Committee. A committee was appointed to draw up resolutions regarding Standard Oil's Schoper Mine northeast of Carlinville. The resolution was to urge the company to open this mine at its earliest possible date.

ℵ Thomas Williamson was in Carlinville **March 14, 1928,** conducting business for Standard.

Car Stolen In Standard Addition

ℵ Jack Leach, chef at Ariston Restaurant in Litchfield, IL, parked his car in Standard Addition on **May 16, 1928.** While he was in the store in Standard Addition, someone drove away in his car and no trace of it was ever found.

ℵ C.W. Clark went to Wood River **September 12, 1928,** on business for Standard Oil.

Owners of Exploded Store Awarded Damages

ℵ On **December 5, 1928,** Ciaccio and Scaperia, owners of Macoupin Grocery Store that exploded in Standard Addition in **August of 1926,** were awarded $25,000.00 from insurance companies, with whom they had insurance at the time of the explosion.

ℵ The **March 13, 1929, Carlinville Democrat** said, *"The result of the election for control of Standard Oil of Indiana is accepted by many in Carlinville, as a hopeful sign that the long closed coal mines owned by that company near this city, may be reopened. It was known that Col. Stewart, chairman of the board, has never favored the policy of operating coal mines. Citizens of Carlinville who have personally appeared before him and petitioned for reopening of the* *mines here have been most courteously treated, but have come away with the conviction that Col. Stewart believed that Standard should not mine coal. Now that he has been deposed and his successor appointed, there is hope that the new administration may reopen the mines here. It is known that some men on the board of directors are in favor of such action. No one in authority has given any statement, but it would seem to be an opportune moment for the Chamber of Commerce to renew its efforts to induce the company to reopen the mines. It might result in much good if an appeal is made to the new management."*

On **June 5, 1929,** at 1:00 p.m., the large brick building of Standard Oil, located on the site of the old Carlinville Mine property was destroyed by fire. Martin and Walter Jones had been using it as a plant for wrecking cars and storage of parts. The origin of the blaze was unknown.

Former Schoper Employee Takes Life

On **September 25, 1929,** the Carlinville Democrat reported, *"A former employee of Schoper, who had moved to Chicago with his wife, shot his wife at their Chicago home and then turned the gun upon himself. Their remains were brought to Carlinville for burial."*

The **October 9, 1929**, Carlinville Democrat didn't have a much better report, when it stated *"Family troubles and liquor were the cause of tragedy, which took place at Standard City, at 5:30 p.m., when a man took his own life by cutting his throat. The man owned a store at Standard City. Leading to the trouble was a suit filed by his wife for divorce."*

The **December 18, 1929,** issue of the Carlinville Democrat reported the mining situation was quiet in Macoupin County. It went on to state, *"There seems to be a conflict between United Mine Workers of America and the National Union. Conflicts have arisen between the two forces in Christian and Sangamon Counties and now Macoupin. Some disorder was threatened in Staunton, Mt. Olive and Gillespie mines this week, but there was no violence and the mines are in operation with the usual miners at work."*

Elmer Pocklington, 18, son of Mr. and Mrs. Frank Pocklington, broke his ankle skating on the reservoir at Schoper on **February 19, 1930.**

Standard City Holds Election

On **April 23, 1930,** Standard City held an election. Those put into office were, John Krivee – Village President; Joe Begando – Village Clerk; Leo F. Bloome – Village Treasurer. John Chiappini, Tony Bogetto, Louis Mussatto and Frank Firek were all elected Village Trustees.

Standard paid their tax bill on **April 30, 1930**. The check was for $24,443.60 and was for all Standard's property located in Macoupin County.

Mr. and Mrs. R. Fitzgerald of Whiting, IN, were guests of friends in Carlinville on **October 8, 1930.** The Fitzgerald's are formerly from Carlinville and he worked for Standard Oil here.

Items of Interest Picked Up Here and There By J.E. McClure

On **December 10, 1930**, the Carlinville Democrat's "Items of Interest Picked Up Here and There" column by J.E. McClure had a lot to say about Standard Oil. Mr. McClure was talking about questions he was frequently asked, *"When did Standard Oil Company come to Carlinville? How long have the coal mines been closed?"* He said that he had been asked these two questions again recently and thought perhaps the readers could answer both of them promptly, but he could not. He went on to say, *"I had to brush the cobwebs out of my memory and try to fix upon the year when the great business activities came to Carlinville and I couldn't do it definitely."* He asked W.G. Bartels about it and, of course, he knew. So when did Standard Oil Company come to Carlinville? Read on.

Note: What follows is from Mr. McClure's perspective.

"I believe that I can find a purchaser for your coal mine," said A.W. Crawford to W.G. Bartels. Mr. Bartels was a successful coal man. He knew the business from top to bottom. He had been in it since he was 15. He helped his father operate a coal mine when coal was hoisted by horsepower. He had a hard time getting his father to help him when he sunk a new mine. He wanted modern equipment. He bought engines and boilers at a bargain, but they had to be paid for before they were shipped. The father yet believed that horses could furnish sufficient hoisting power. The son knew better and didn't want to give up. He told his troubles to James K. Furber at Anderson Bank. *"How much money do you need?"* said Mr. Furber. *"I have to have $4,000."* *"All right,"* said Mr. Furber, *"You can have it."* Years later, Mr. Bartels took over the coal mine in the north part of Carlinville and formed Carlinville Coal Company. It was a prosperous venture. He and his associates enjoyed substantial dividends.

Mr. Bartels thought seriously about Mr. Crawford's proposal. It was a good time to sell. Experts came and went through the mine. They inspected all parts of it, all of which Mr. Bartels had completely modernized. The reports concerning the mine were favorable. Mr. Crawford and Mr. Bartels negotiated and a price was agreed upon. All rights, title and interest in the Carlinville Coal Company were sold and transferred, but to whom? The deed said to Thomas Williamson as trustee. Mr. Williamson was not a coal operator, he was a lawyer. Carlinville's curiosity was quite consuming. We suspected that he was representing Standard Oil Company. *"Why Standard Oil Company?"* was the question on everyone's lips. Nobody knew. I learned later that that a few did, but they did not tell. How splendid it would be if our coal fields could be developed by a great organization, which never fails. Local dreamers and idealists were indulging in their *"I told you so's."*

In the meantime, Mr. Crawford was busy buying coal rights and Mr. Williamson was busy passing titles, and accepting them as trustee. Then came an announcement, Herman Schoper sold his farm of 535-acres in Macoupin County and grantee was "Thomas Williamson, Trustee." Then it was revealed that 23,000-acres of coal lands had been purchased. With the Schoper land, they had more than a 1,000-acres of surface in eastern

Macoupin County. We began to know and understand. Mr. Bartels retired from the coal company. E.M. Clark became superintendent of activities here. Frank R. Fisher, an experienced coal man, was placed in charge of Carlinville Coal Company. Vast improvements were set in motion. Who was Mr. Clark? He was actively and prominently identified with Standard Oil. Yes, it was Standard Oil Company of Indiana who had come among us. Things never dreamed of were coming to pass. Standard Oil was going to sink the greatest and best-equipped coal mines in the world. This was going to be done on the very doorstep of Carlinville, and they lost no time in doing it. Plans were made, projects undertaken, which to our limited vision seemed stupendous and almost incomprehensible, but to Standard Oil, they represented another carefully conceived undertaking, which would add to the prestige and profit of that great organization. *"Why did they select Carlinville and this section of Macoupin County?"* was thought by hundreds of us. An official of Standard Oil said, *"The company has faith in this locality and that faith has largely been influential in the purchase of the mine and its adjoining properties."*

Events occurred so rapidly people could scarcely keep up with them. Every train brought people to Carlinville. St. George Hotel was filled every night. Traveling men learned that they must telegraph ahead for reservations. Civil and mechanical engineers came and took up their abode in Carlinville. They had months, perhaps years of work ahead of them. Additional miners were employed at the mines everyday. Soon there were no vacant houses in the city. Something had to be done about it. I recall how perfectly splendid cooperation between the people of Carlinville and Standard Oil was. *"Build houses,"* said a representative of the company. *"Everyday will add to the population of Carlinville. Ultimately 2 great mines, possibly 3 will be sunk. Ultimately 2,000 men will be employed, more than 1,200 cars of coal will be hoisted every week."* Standard was positive about a number of things. They emphasized the fact that they would employ only high-class men; that they would mine no coal for market. Therefore, the mines would be kept in constant operation, producing fuel required for three refineries. They wanted their employees to have every opportunity and every advantage that every other citizen enjoyed. First they must have good homes in which to live. They wanted these homes to be within and a part of Carlinville. They did not want isolated settlements of coal miners. So there must be additions to Carlinville. Water mains must be laid; streets must be paved; sewers must be built; sidewalks must be constructed and adequate and complete arrangements for education of the children. All of which presented a big municipal program, a realization of an ideal. A greater Carlinville was coming with the morrow. The leaders and chiefs in the great enterprise, who had come to Carlinville, were big men of big mind and big heart, representing and speaking for a big corporation. It was a great pleasure to carry on with such clean-cut broad-minded men. *"What can we do to help you?"* was always answered by any representative of Standard Oil to whom it was propounded, *"Build, build, and prepare to take care of our people."* Carlinville set about the task and worked at it right manfully.

The St. George Hotel in early ***1900's.***

A stupendous building program was undertaken at Schoper. The workmen lived in Carlinville and the company wanted all of them to live in the county seat. The connecting road was in poor condition. The company felt that it should be put in such condition that it would be good for travel all months and days of the year. They would furnish cinders. Did Carlinville want to assume responsibility of building the all-weather road? A citizens meeting followed. The cost would be $4,000. It was raised within a few days and the road was built. We were mighty glad that great prosperity for us was around the corner. We tried in every way to show appreciation. It was a new experience for us to so suddenly be confronted with, the advent of an institution, which would double our population. I recall how the very magnitude of the whole thing seemed to overwhelm us. Was it really true, or were we awakening from a dream? And came the answer daily from the courteous, business-like officials who were directing the big installation, *"build, build houses, and take care of our people."*

Those were great times. I have thought about them, read about them and talked about them within the last few days. I can even yet experience the thrill we all enjoyed in those days. I almost forgot to tell when it was. Carlinville Coal Company was sold to Standard Oil in **October of 1917**. What I have told took place between that time and **February of 1918.** Those were big and epochal times in the history of Carlinville. Other communities have had booms. Thousands of them. But none like Carlinville.

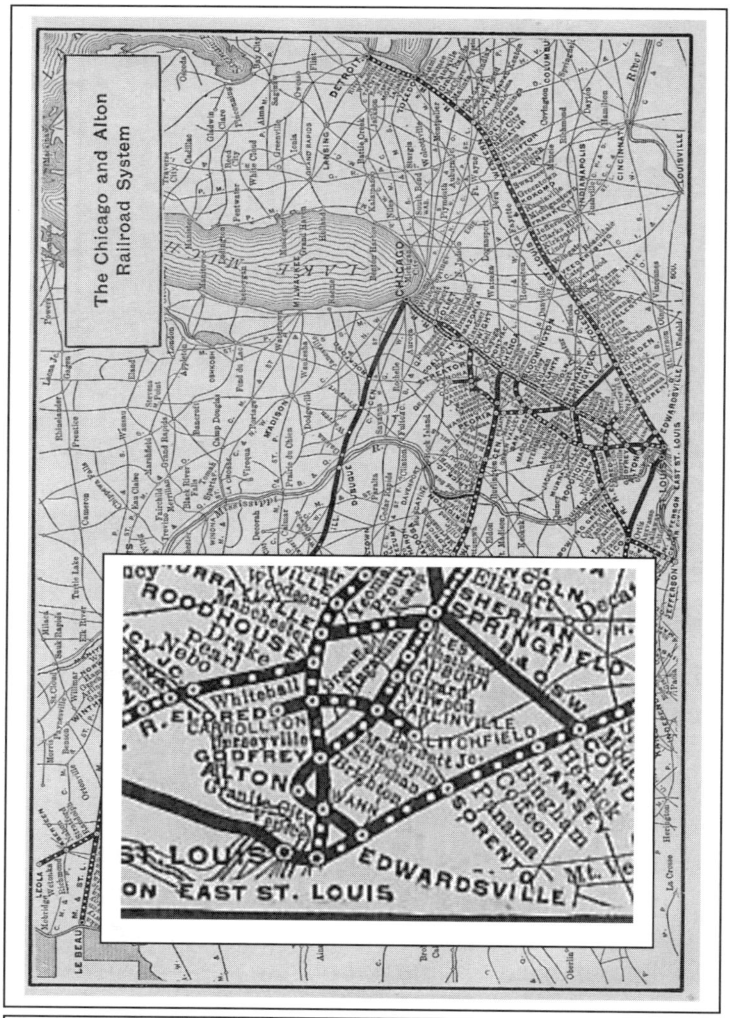

1928 Chicago & Alton Railroad Map.

Note: Mr. McClure summed up the feeling and mood of those times well. Standard Oil made a promise, but did not keep their promise. The mines did not stay open constantly. They first closed Carlinville Coal Mine, then Berry was closed, supposedly temporarily, and it never reopened even though Schoper was up and running periodically for the next couple of years that followed. Finally, Schoper closed. If you read the articles earlier in this book, you will see that Schoper (as well as Berry, for the time that it was open) didn't have too many months where they worked steadily until a strike or a shutdown took place; sometimes it was four to six months before the mine would reopen. The last shutdown of Schoper was only supposed to be temporary, but it turned out to be permanent. Standard knew this was permanent from the time they closed it; but just like in the beginning, they kept the people it mattered most to – the people of Carlinville – in the dark. Maybe Carlinville would have been better off if Standard hadn't come in the 1st place, but they did come and Carlinville was left with it all.

☙ An announcement on **December 10, 1930,** stated C & A Railroad and all of its property was to be sold at auction the next afternoon by a master in chancery at Wilmington, west of Chicago. The C & A had been in financial straits for years. Carlinville people thought of the railroad company as "*Their own beloved C & A. We folks here in Macoupin County have a certain sentiment for it, but it is going to be sold. The purchaser will not pick it up and carry it away, it will continue to be with us.*"

The purchaser was Baltimore & Ohio Railroad Company. Probabilities were assumed that the C & A would be greatly improved and made even better than it was. It was a tough time for railroads; everybody owned automobiles by this time. The C & A came to Macoupin County in **1850.** It was a great force for development of the county, and now was going to pass into other hands. Officials and methods as well as the name were changed, but it still was the *"old C & A"* to people in the area and in Carlinville.

Items of Interest Picked Up Here and There - By J.E. McClure

On **December 17, 1930**, the Carlinville Democrat's "Items of Interest Picked Up Here and There" column by J.E. McClure had more to say about Standard and Carlinville.

Note: This is told from Mr. McClure's perspective.

"I was delighted with your Items of Interest last week. I don't know when I've enjoyed your column as much as I did reading of the boom days when Standard came to Carlinville," said a friend of mine. Then he continued, *"Those were exciting times here; we were enjoying all of the thrills of a boom town."* Then he and I had fun reminiscing. We recalled the events of those years. We had never been in a town where they struck gold, but we were living in a town where they struck coal and lots of it. In back of the huge undertaking was Standard Oil. There was a boom, but it was accompanied by no uncertainty. Could our people meet the big demands made upon them? That was the only cloud in the sky. I recall a special occasion when our citizens sought to show appreciation. The Commercial Club was the medium. A banquet was held. Representatives of Standard Oil in Carlinville were guests of honor. We broke bread together and touched elbows. Many of us had become personally acquainted with Mr. E.M. Clark, the manager. He is now one of the big men of Standard Oil in this country. His office is in New York City. At that banquet there were many speeches to be made. Mr. Clark was unable to be present so Mr. Williamson, company attorney, spoke for him. I recall the inspiring address he delivered. His heart was in what he said. He was born in Macoupin County, he was glad to see this community so highly favored. He visualized the future and as he did, told of the greatness, which came to Wood River when Standard Oil went there, and built and built, until that great oil center is known the world over. All of the representatives of Standard Oil who spoke at that banquet expressed pleasure at the reception they had received and were delighted at the manner in which Carlinville was doing its part. One of them referred to the fact that our people must be up and doing in order to accomplish that particular thing which the company so much desired: To provide sufficient housing so their employees might all reside in Carlinville. Other communities were seeking to attract them. We must meet the demand. The more we heard from those men, the more we were impressed with the magnitude of the whole undertaking. Captain Rice asserted that when the 2 mines were sunk and in full operation, they would hoist 10,000 tons of coal a day. In 3 or 4-years we would have a population of 8-10 thousand people. Conditional, of course, that Carlinville could take care of them.

There was a lot of work done by our citizens. Mr. Clark was a source of inspiration to everyone. A group of representative citizens met one day in conference. I was in the party. There were no useless preliminaries. We came to a point where they would enlarge the city and build houses. It was the sense of that meeting that all of our people should be urged to cooperate in a house building program. How much money would be ample to begin operations? S20,000.00 was subscribed on the spot. We went from that meeting encouraged, inspired and enthusiastic. Out of it all came the Carlinville Home Builders Association. Shares were sold at $25.00 each. Many of our citizens became subscribers to the stock. Contracts were let and houses were built. I don't know how many houses were sold or what became of them, but I do know that that body of men devoted their time, spent their money and worked hard and accomplished a lot. Carlinville was putting forth a supreme effort. The city council added 175-acres in five subdivisions into the city. That added $100,000 to real estate property values within the city of Carlinville. Public improvements were constructed at a rapid rate. Additions must have good road connections with schools; business sections of the city and with the cemetery, so sewers, pavements, walks were underway. In the mean time the slogan continued, *"build, build houses."*

Will Otwell owned land adjoining the city. It became part of Carlinville by vote of the people. He built many houses and invested large sums of money. The Standard Oil Addition, the Arnett & Murphy Addition, the Crawford & Ibbetson Addition, the Burke Addition, were all soon dotted with homes, and yet this wasn't enough, Standard was moving along with its carefully wrought plan and keeping ahead of the house builders. Everyday brought new residents. It was a rapid and stable, growth in population, in financial resources; and permanent. But, we just couldn't keep up with the demands. We were near our wits end and then Standard threw itself into the balance. They ordered and quickly built 156 two-story, modern homes. They were all located in Standard Addition on what had been 40-acres of farming land. The erection of those houses relieved the strain and enabled the home builders, the time they needed to do their part in house construction. New additions called for new sewers, paved streets, school buildings, more places of business, a general enlargement of all lines of activity and business to meet the requirements of an ever-increasing population. Throughout it all, those men who spoke for Standard were sincere, honest, and truthful. They believed everything they said, just as we believed it. There was no deception. I knew that every man who had a part in it was sincere. Each saw Carlinville as the center of a great coal mining industry, which would produce all the fuel needed by the three great oil refineries of Standard Oil, and said so. Those were happy and memorable times for us all. Scarcely a man or a woman, but that enjoyed the prosperity, which came with Standard Oil. That company spent millions. It had 23,000-acres of coal lands.

☙ **Items of Interest Picked Up Here and There By J.E. McClure** ❧

☙ On **December 31, 1930**, the *Carlinville Democrat's* "Items of Interest Picked Up Here and There" column by J.E. McClure had even more to say about Standard and Carlinville.

Note: This is told from Mr. McClure's perspective.

"I wish you would continue reminiscing about Carlinville's boom days," said a Carlinville citizen to me. *"Those extraordinary things happened here among us, within the last 12-years, but you have refreshed my memory with reference to so many details."* Another reader said, *"I bought, I built, I enlarged my business and profited greatly during the boom days. There was no*

doubt as to its permanency. I have never known another boom like it." My friend was quite correct. We simply had become recipients of good fortune. Standard Oil had availed itself of the coal, which underlies this section of the county. They were going to dig their own fuel here for their refineries, which would not close down. Neither would the mines. No wonder we thought ourselves fortunate. Our population grew. Sewers were built, water mains extended, walks constructed and pavements laid. Carlinville was doing its utmost. Store buildings were enlarged and new ones built. New lawyers, new doctors, new contractors were identifying themselves with Carlinville They needed homes and offices. Those were great times.

Standard told us they would develop this great coal field by means of the best equipment procurable. And they did. Visitors came 1,000's of miles to see what Standard had done here. They went away feeling that Berry and Schoper Mines were in a class by themselves, so perfect were they. The 1,300 miners who were working in those mines were hoisting an unprecedented amount of coal. That coal was crushed and sent away in long trains of coal cars owned by Standard Oil, each having on its side in big bold letters, *"Not to be taken off of the C & A Railroad."* New men were hired every day. The company selected its employees with great care. A representative of Standard Oil said to me, *"We want to employ most efficient workmen, who will be great assets to Carlinville when they become permanent residents here."* Day by day men were put to work. Employee's families came shortly after and a home awaited them. They liked Carlinville and we liked them. They attended our churches, their children enrolled in our schools, they patronized our stores, they joined our lodges, and they entered into our civic activities. They quickly became a part of us. They were mighty fortunate coal miners to secure permanent employment with such a company.

Then one day in **1924**, there was a notice posted on one of the mines,*"This mine will be closed indefinitely within three days."* That came as a thunderbolt out of the clear blue sky. It shocked and took the breath of everyone. Soon a similar notice was posted at the other mine. Committees of miners who sought information, learned little because the order had come from headquarters. The Chamber of Commerce wrote. Committees of citizens visited, but the mines remained closed. The explanation was this, *"The mining industry is greatly overdone. There is a surplus of production. We can now buy screenings at less than we can put the uncrushed coal on the surface. So long as this condition obtains it is folly for us to mine coal."* And that was the truth. They did not misrepresent anything to us. A condition had arisen, which their wisest men had not foreseen. If they had anticipated such a condition they would not have invested millions here, in and around Carlinville. I have always understood that Col. Stewart, who was then head of Standard Oil of Indiana, was opposed to the company engaging in coal mining. I have been told that he stood steadfastly against it, but was overruled by his associates. When the great break in coal came, Col. Stewart felt that he had been vindicated. His *"I told you so"* found expression in the orders to close the mines. The closing of the mines was but a relatively small transaction with one of the world's greatest business organizations, but it was a black Friday for Carlinville and 1,300 coal miners. Those miners began seeking employment elsewhere. Their families must have food and clothing. They began to move gradually and then rapidly. Many sacrificed what they had paid on their homes. There was a good many tears shed in those days. I think that now there are 100 miners who yet live in Carlinville and work elsewhere. Carlinville felt the blow keenly. Since those orders were posted, store after store has closed. Professional and businessmen have left. Offices have been vacated. There are more vacant storerooms and residences here now, than ever before in the history of Carlinville. Arc lights illuminate unoccupied houses and unused streets. Carlinville's population is now barely 4,000; at the time the mine shut down it was upward of 6,000. The people of Carlinville had accepted the company's insistent appeal, *"build, build houses, and take care of our people,"* in good faith. They had given no thought to the possibility of a closing down order. It took Carlinville by complete surprise. Standard is no longer functioning in this community. A few remain to look after closed mines and other interests, but there are no activities. It may be the height of presumption for me to make such an assertion, but somehow I feel Standard Oil owes this community a moral obligation. Somehow, I believe Standard Oil will sometime, come back to Carlinville and break the seals on the mines.

It was announced on **December 31, 1930**, that Standard Oil of Indiana would sue Standard Oil of Colorado. The suit was to determine if Standard Oil of Colorado could legally use the words "Standard Oil" in its name. Standard Oil of Colorado was in no way associated or affiliated with Standard Oil of Indiana.

Items of Interest Picked Up Here and There By J.E. McClure

On **January 7, 1932**, the Carlinville Democrat's "Items of Interest Picked Up Here and There" column by J.E. McClure remembers the boom times of Standard and Carlinville.

Note: This is told from Mr. McClure's perspective.

In recent issues of this paper, I have told my recollections of Standard Oil coming to this community. When they came, we all felt proud. We didn't know why that great institution had singled out Carlinville as the place where they would spend millions. We subsequently learned that it was our great mineral resources, our railroad, the quality of our citizenship and other advantages which induced that great industrial force to come here, backed with their great power and their great capital. They came, they bought, and they established themselves here. We were elated. Our people needed no encouraging or urging because our sense of appreciation prompted us to go to Standard and ask, *"What can we do to help?"* The answer to that was, *"Add new additions to your city, pave your streets, construct sewers, build schoolhouses, extend your water mains, build houses, take care of our people."*

One day my friend W.B. Otwell told me that he was on his way to the Standard Oil office in Carlinville, to talk with Mr. E. Clark. When the great industrial enterprise came to Carlinville, he came as its directing head. He handled men and finances with highest admiration. He preached common sense, reciprocity, humanity, and good wages for good work and good homes for good employees. He told us how to make our city beautiful, rich and prosperous and among other things he told us, *"Come and go along with the great Standard Oil Company for they intend to run every day for 50-years. Standard Oil is going to pay such wages that men have good homes with modern conveniences."*

Mr. Otwell told me that he would come back and tell me of his talk with Mr. Clark. He did so. *"Is there anything that a common citizen can do to help?"* inquired Mr. Otwell. Mr. Clark expressed his vision of the great project, and then said, *"You own some very beautiful suburban acreage. We would have you put in sewers and sidewalks and plant trees. Then we would have you cover it with houses so that our miners would have good, comfortable homes." "But,"* inquired Mr. Otwell further, *"suppose Mr. Clark that we citizens of Carlinville spend our money in the line indicated by you and you should close up business and move away and cease to operate your great plants, what would become of us?"* Mr. Clark's answer was very significant and Mr. Otwell said he would never forget it. Mr. Clark replied, *"Did you ever know Standard Oil Company to fail on a proposition where they had started?"* Mr. Otwell did not, so as far as he was concerned, he believed in them. Mr. Otwell was peppier than I had ever seen him. He was going to undertake stupendous things, he was going to do as Mr. Clark recommended. He went to the lumberyard and to carpenter shops and before long 25 carpenters and many masons, teamsters and ditch diggers were busily engaged in Otwell Addition. As a result of that great activity, six different Otwell Additions were added to Carlinville. They are all yet within the city limits. 250 lots were laid out. Sewers constructed, sidewalks built, and trees planted. They could not build houses fast enough for miners who wanted homes. Mr. Clark and Mr. Otwell became good friends. One day Mr. Clark drove through the Otwell Additions. Mr. Otwell was heels over head in building, painting, digging and planting. Mr. Clark left his automobile and shook hands with Mr. Otwell and congratulated him. *"You are doing fine, but you are not building the houses big enough,"* remarked Mr. Clark. Mr. Otwell's reply was honest and full of logic, *"I know how to do little things, but I never learned how to do big things. Get somebody else to build the big houses and I will build the little ones."* Mr. Otwell was very proud of the additions and the people who lived in them. He was talking to me recently and said, *"Jim, I think you are right and that there is some moral obligation resting on the shoulders of the great Standard Oil with reference to Carlinville. Just by the turn of their hand they could bring back prosperity where prosperity deserves to be. They could restore to 1,200 men and their families something that has dropped out of their lives."* I was convinced that others were thinking this same way. What Will Otwell did in those days, Earnest Ibbetson, Arnett & Murphy, the Home Builders Association and the promoters of Stratton Addition and so many of others did to a greater or lesser degree. Now, most of those homes are empty, but the tax collector's notice will come in a few weeks, just as usual. The story of Carlinville's boom days has extended over several weeks, maybe I have carried it on too long, but it had never been told before. Maybe somewhere these stories will be read by those in authority and they will prove to the world that Standard has not, in fact, failed Carlinville.

Schoper Water Comes To Aid of Carlinville

The **January 21, 1932,** Carlinville Democrat reported that, *"Through cooperation of officials of Standard Oil, water at Schoper Lake, was turned into Macoupin Creek in order to replenish the supply for the reservoir south of town, which supplies this city with water. Byron Graham, superintendent, and C.W. Clark, manager, were at Schoper when the water was released. Just what the result of this will be is problematical. It is 14-miles from Schoper to the creek through the country. The temperature*

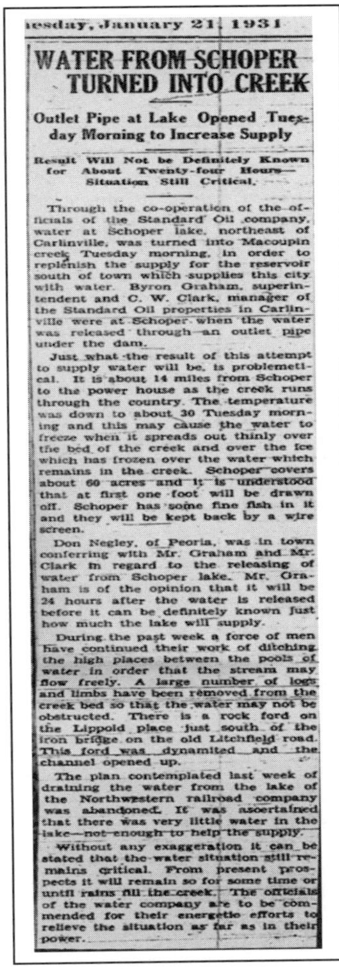

...esday, January 21, 1931

WATER FROM SCHOPER TURNED INTO CREEK

Outlet Pipe at Lake Opeped Tuesday Morning to Increase Supply

Result Will Not be Definitely Known for About Twenty-four Hours—Situation Still Critical.

Through the co-operation of the officials of the Standard Oil company, water at Schoper lake, northeast of Carlinville, was turned into Macoupin creek Tuesday morning, in order to replenish the supply for the reservoir south of town which supplies this city with water. Byron Graham, superintendent and C. W. Clark, manager of the Standard Oil properties in Carlinville were at Schoper when the water was released through an outlet pipe under the dam.

Just what the result of this attempt to supply water will be, is problemetical. It is about 14 miles from Schoper to the power house as the creek runs through the country. The temperature was down to about 30 Tuesday morning and this may cause the water to freeze when it spreads out thinly over the bed of the creek and over the ice which has frozen over the water which remains in the creek. Schoper covers about 60 acres and it is understood that at first one foot will be drawn off. Schoper has some fine fish in it and they will be kept back by a wire screen.

Don Negley, of Peoria, was in town conferring with Mr. Graham and Mr. Clark in regard to the releasing of water from Schoper lake. Mr. Graham is of the opinion that it will be 24 hours after the water is released before it can be definitely known just how much the lake will supply.

During the past week a force of men have continued their work of ditching the high places between the pools of water in order that the stream may flow freely. A large number of logs and limbs have been removed from the creek bed so that the water may not be obstructed. There is a rock ford on the Lippold place just south of the iron bridge on the old Litchfield road. This ford was dynamited and the channel opened up.

The plan contemplated last week of draining the water from the lake of the Northwestern railroad company was abandoned. It was asoertained that there was very little water in the lake—not enough to help the supply.

Without any exaggeration it can be stated that the water situation still remains critical. From present prospects it will remain so for some time or until rains fill the creek. The officials of the water company are to be commended for their energetic efforts to relieve the situation as far as in their power.

was down to 30 degrees and may cause the water to freeze when it spreads out thinly over the bed of the creek and over the ice, which has frozen over the water that remains in the creek. Schoper Lake covers 60-acres and at first, 1-foot will be drawn off. The fish in Schoper will be kept back by a wire screen. It will be 24 hours after water is released before it will be known just how much water the lake will supply. During the past week, a force of men have continued their work of ditching high places between the pools of water in order that the stream may flow freely. A large number of logs and limbs have been removed from the creek bed, so the water won't be obstructed. Without exaggeration, it can be said, the water situation remains critical."

☙ On **January 28, 1932,** it was reported the *"Water situation was much improved. The water from Schoper Lake had reached the pumping station. It took 4 days to travel 14-miles. The critical situation regarding water for Carlinville has been relieved for now. The outlet on Schoper Lake was opened on* ***January 20th.*** *The thermometer was 10 above zero and this retarded the flow. However, the weather moderated the next few days, which brought results, and water reached the reservoir in front of the pumping station. Officials of the water company estimated, that between 7 and 8 million gallons of water were released from Schoper. The rise at the pumping station was noticed Friday afternoon. Saturday afternoon there was considerable increase in depth. By Sunday, it had risen to a depth of 15-inches from no depth at all directly in front of the pump house. The distance the water traveled was 14-miles. It took 4 days for it to reach the dam at the reservoir."*

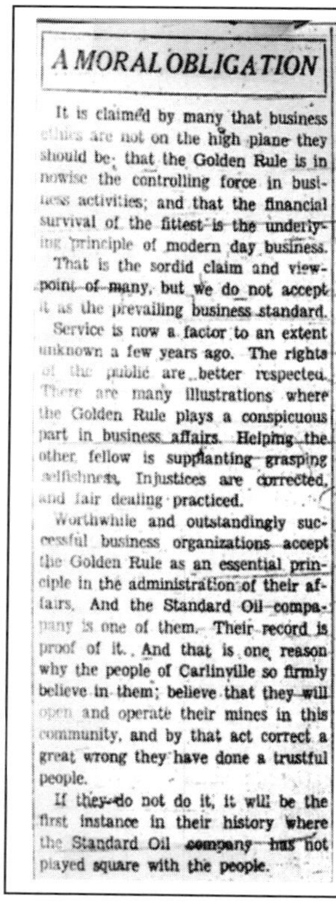

A MORAL OBLIGATION

It is claimed by many that business ethics are not on the high plane they should be; that the Golden Rule is in nowise the controlling force in business activities; and that the financial survival of the fittest is the underlying principle of modern day business.

That is the sordid claim and viewpoint of many, but we do not accept it as the prevailing business standard. Service is now a factor to an extent unknown a few years ago. The rights of the public are better respected. There are many illustrations where the Golden Rule plays a conspicuous part in business affairs. Helping the other fellow is supplanting grasping selfishness. Injustices are corrected, and fair dealing practiced.

Worthwhile and outstandingly successful business organizations accept the Golden Rule as an essential principle in the administration of their affairs. And the Standard Oil company is one of them. Their record is proof of it. And that is one reason why the people of Carlinville so firmly believe in them; believe that they will open and operate their mines in this community, and by that act correct a great wrong they have done a trustful people.

If they do not do it, it will be the first instance in their history where the Standard Oil company has not played square with the people.

☙ A Moral Obligation ❧

☙ The **January 28, 1932, Carlinville Democrat** stated, *"Worthwhile and successful business organizations accept the Golden Rule as an essential principle in administrating their affairs. Standard Oil is one of them. Their record is proof of it, and that is one reason why people of Carlinville believe so firmly in them; believe that they will open their mines in this community, and by doing that act, correct a great wrong that they have done a trustful people. If they do not, it will be the first instance in their history where Standard Oil Company has not played square with people."*

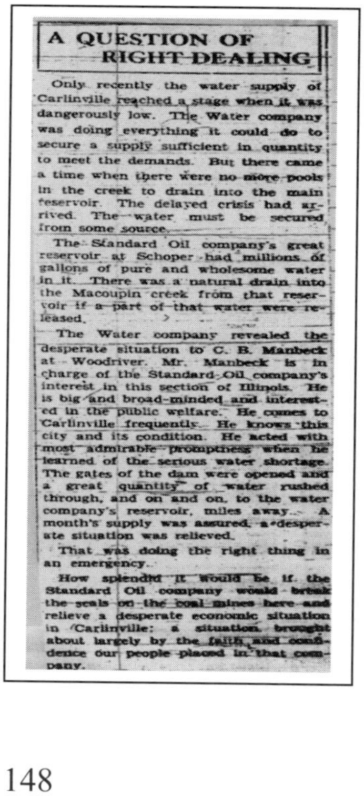

A QUESTION OF RIGHT DEALING

Only recently the water supply of Carlinville reached a stage when it was dangerously low. The Water company was doing everything it could do to secure a supply sufficient in quantity to meet the demands. But there came a time when there were no more pools in the creek to drain into the main reservoir. The delayed crisis had arrived. The water must be secured from some source.

The Standard Oil company's great reservoir at Schoper had millions of gallons of pure and wholesome water in it. There was a natural drain into the Macoupin creek from that reservoir if a part of that water were released.

The Water company revealed the desperate situation to C. B. Manbeck at Woodriver. Mr. Manbeck is in charge of the Standard Oil company's interest in this section of Illinois. He is big and broad-minded and interested in the public welfare. He comes to Carlinville frequently. He knows this city and its condition. He acted with most admirable promptness when he learned of the serious water shortage. The gates of the dam were opened and a great quantity of water rushed through, and on and on, to the water company's reservoir, miles away. A month's supply was assured, a desperate situation was relieved.

That was doing the right thing in an emergency.

How splendid it would be if the Standard Oil company would break the seals on the coal mines here and relieve a desperate economic situation in Carlinville; a situation brought about largely by the faith and confidence our people placed in that company.

☙ Question of Right Dealing ❧

☙ The **February 4, 1932, Carlinville Democrat** stated even more, *"Only recently, the water supply of Carlinville reached a critical stage. Standard Oil's great reservoir at Schoper had millions of gallons of water in it. The water company revealed the desperate situation to C.B. Manbeck at Wood River, who is in charge of Standard Oil's interests in this section of Illinois.* He acted with most admirable promptness when he learned of the serious water shortage. The gates of the dam were opened and a great quantity of water rushed on and on, to the water company's reservoir, miles away. That was doing the right thing in an emergency. How splendid it would be if Standard would break the seals on the mines here and relieve a desperate economic situation in Carlinville, a situation brought about largely by the faith and confidence our people placed in them.

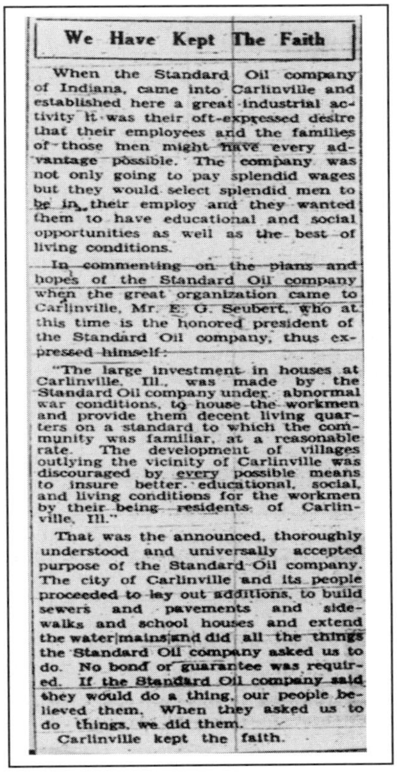

We Have Kept The Faith

When the Standard Oil company of Indiana, came into Carlinville and established here a great industrial activity it was their oft-expressed desire that their employees and the families of those men might have every advantage possible. The company was not only going to pay splendid wages but they would select splendid men to be in their employ and they wanted them to have educational and social opportunities as well as the best of living conditions.

In commenting on the plans and hopes of the Standard Oil company when the great organization came to Carlinville, Mr. E. G. Seubert, who at this time is the honored president of the Standard Oil company, thus expressed himself:

"The large investment in houses at Carlinville, Ill., was made by the Standard Oil company under abnormal war conditions, to house the workmen and provide them decent living quarters on a standard to which the community was familiar, at a reasonable rate. The development of villages outlying the vicinity of Carlinville was discouraged by every possible means to insure better educational, social, and living conditions for the workmen by their being residents of Carlinville, Ill."

That was the announced, thoroughly understood and universally accepted purpose of the Standard Oil company. The city of Carlinville and its people proceeded to lay out additions, to build sewers and pavements and sidewalks and school houses and extend the water mains and did all the things the Standard Oil company asked us to do. No bond or guarantee was required. If the Standard Oil company said they would do a thing, our people believed them. When they asked us to do things, we did them.

Carlinville kept the faith.

☙ We Have Kept the Faith ❧

☙ The *Carlinville Democrat* continued with its attempt to get Standard to reopen the mines, in its **February 11, 1932** issue, *"When Standard came to Carlinville and established here a great industrial activity, it was their oft expressed desire that their employees and their families have every advantage possible. The company was not only going to pay splendid wages, but would select splendid men to be in their employ and they wanted them to have educational and social opportunities as well as the best of living conditions." In commenting on the plans and hopes of Standard when they came to Carlinville, Mr. E.G. Seubert, president of the company, thus expressed himself: 'The large investment in houses at Carlinville was made by the company under abnormal war conditions, to house the workmen and provide them with decent living quarters on a standard to which the community was familiar. The development of villages outlying the vicinity of Carlinville was discouraged by every possible means. There were better educational, social and living conditions for the workmen if they were residents of Carlinville.' Carlinville did all of the things Standard asked us to do. If Standard said they would do a thing, our people believed them. Carlinville kept the faith."*

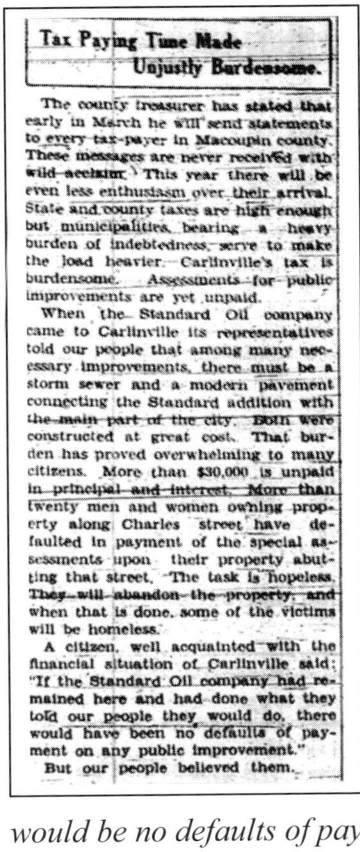

Tax Paying Time Made Unjustly Burdensome.

The county treasurer has stated that early in March he will send statements to every tax-payer in Macoupin county. These messages are never received with wild acclaim. This year there will be even less enthusiasm over their arrival. State and county taxes are high enough but municipalities bearing a heavy burden of indebtedness, serve to make the load heavier. Carlinville's tax is burdensome. Assessments for public improvements are yet unpaid.

When the Standard Oil company came to Carlinville its representatives told our people that among many necessary improvements, there must be a storm sewer and a modern pavement connecting the Standard addition with the main part of the city. Both were constructed at great cost. That burden has proved overwhelming to many citizens. More than $30,000 is unpaid in principal and interest. More than twenty men and women owning property along Charles street have defaulted in payment of the special assessments upon their property abutting that street. The task is hopeless. They will abandon the property, and when that is done, some of the victims will be homeless.

A citizen, well acquainted with the financial situation of Carlinville said: "If the Standard Oil company had remained here and had done what they told our people they would do, there would have been no defaults of payment on any public improvement."

But our people believed them.

☙ Tax Paying Time Made Unjustly Burdensome ❧

☙ The **February 18, 1932**, *Carlinville Democrat* said what every citizen in the county was thinking, *"The County treasurer has stated that in March he will send statements to all taxpayers in Macoupin County. Carlinville's tax is burdensome. Assessments for public improvements are unpaid.*

When Standard came to Carlinville, its representatives told our people that there must be a storm sewer and a modern pavement connecting Standard Addition with the main part of the city. Both were constructed at great cost. That burden has proved overwhelming to citizens. A citizen well acquainted with the financial situation of Carlinville said: 'If Standard had remained here and done what they told our people they would do, there would be no defaults of payment on any public improvement. But, our people believed in them.'"

Would anything come from this one paper fight to get Standard to come to Carlinville and reopen the mines? Could citizens of this town "guilt" the company into it? The great Standard Oil did and does owe Carlinville a "moral obligation." It was said that the money the company spent in Carlinville and the community was just a drop in the bucket to them. If that was the case, why didn't they reopen the mines just to save face – their own face? I am sure that they could have reopened the mines and sold the coal on the open market and made money. If they could buy coal cheaper than mine it, why didn't they take that situation a step further and reopen the mines by leasing or selling them at a great discount on the condition that the lessee had to sell some of the coal from those mines at a great discount to Standard Oil. There had to have been something they could do. Just letting everything fall down around the citizens of Carlinville and the community shouldn't have been the answer. But, what did a great institution like Standard Oil Company care? Nothing – they proved that.

☙ Local Company Sells Mine ❧

☙ **On May 13, 1932,** Carlinville Cooperative Coal Company became new owners of the Carlinville Coal Company Mine in the south part of Carlinville. The price paid was $50,000.00.

☙ Explosion Damages Air Shaft at Mine ❧

☙ At 7:00 p.m. on **December 16, 1931,** an explosion was heard that rocked Carlinville. The airshaft in the south mine, known as Carlinville Coal Mine had been torn loose from its bearings and a hole blown in the brick building that houses the fan. Sheriff Fries and deputies investigated and found 30-feet of fuse, but could not determine what explosive had been used.

☙ Trouble At South Mine ❧

☙ On **February 10, 1932,** 100 miners were stopped from going down into the shaft of the Carlinville Mine. Sheriff Fries and his men went to the scene and a meeting was arranged between union miners and operators. It was hoped a settlement could be reached.

☙ Store and Post Office Robbed ❧

☙ On **March 2, 1932,** the store and post office at Standard City were robbed. Entrance was gained through a rear window. $1.50 in stamps and $1.60 in money were taken.

☙ A meeting of the high school board and the Board of Education was held on **August 17, 1932,** and a contract was signed to bus students from Standard City to Carlinville Community High School. This service was for 1-year and if it proved to be satisfactory it will be continued the following year. They were to use the cinder road from Standard City to Route 4.

☙ Big Storm Strikes Carlinville ❧

☙ The storm, which struck Carlinville on **May 24, 1933,** around 6:00 p.m., started with the rumbling of thunder in the northwest. It increased and the storm broke out at about 7:00 p.m. in Carlinville. At 1st it was wind, then rain began, the wind increased and then it came in torrents.

Strong winds blew down trees all over the city. About 25 trees were uprooted or broken off. Five went down across College Avenue, knocking out electric and telephone service. Mrs. R.E. Crum's residence on Johnson Street was badly damaged with the roof crushed in two places and a big glass window in front broken to pieces. Mrs. Crum was at home alone at the time. It was a shocking experience and she hopes to never have another one like it. One of the strange sights the storm produced was the felling of a monster maple tree on Nicholas Street about 2-blocks west of the college. It had the appearance of two big trees, which had grown together. It was on the south side of the street near the home of Charles Cundall. When this old timer came down, it was with roots and all. North Broad was blocked by a big tree, 100's of limbs were down all over town. Many people went to their doors and looked out on the tempest. They noted a whirling motion of the wind and heard a rumbling high up in the air. It sounded like the roar of a distant train. Not only people in Carlinville heard this noise, but many in Chesterfield heard it as well. Had this rumbling reached the ground before it hit this city – well there could have been a much different story to tell. Fortunately no lives were lost in this one.

☙ Carlinville Bracing For Another Storm ❧

☙ A different kind of storm had been brewing in Carlinville for a long time. It was a storm that in this case, would not have a happy ending. Citizens in and near the city came to realize that Standard Oil Company had come to this area with big promises and big dreams, which people were all too happy to go along with, without even a question. Now Standard Oil was not coming back. They had packed up, sealed up and left without even a thank you for what Carlinville had done to help them or a good bye, because they didn't want to be faced with the outrage if they told citizens the truth.

I believe that Standard knew on the day they closed Schoper for the last time that they were not coming back and the mine would never reopen. To let the mines and houses fall into neglectful states was disrespectful to Carlinville as a whole. It's as if Standard said, *"Incorporate our Addition into your city and then we'll abandon it and you'll have to look at it and take care of it, as we will be miles away in our big offices at our big desks still making our big paychecks."*

As I asked earlier, would Carlinville have been better off if they hadn't come at all, I really wonder?

As to what Standard Oil thinks in regard to Carlinville, Schoper and the area, in his book *"Standard Oil Company – Oil Pioneer Of The Midwest,"* on page 202, Paul H. Giddens said it best, *"Those (Standard Oil Company officials) who recall this experience regard it as an episode better off forgotten."*

The Standard Oil Company had a few things in the Carlinville area left to take care of in those days and then they could leave and forget that they had ever been here.

First they were going to have to end all of their ties with the community and that meant selling the houses at Carlinville and Schoper.

It would be a few more years before Standard would sell the homes, but eventually they did.

Chapter X

"Standard Oil Company Unloads"

Standard Oil Company Unloads

Note: There was not any information or articles related to the Standard Oil Company's activities in or around Carlinville, Schoper or Berry for almost a year and then…

☙ Shortage of Homes To Rent In Carlinville ❧

☙ The **August 22, 1935**, Carlinville Democrat said there was a shortage of homes for rent in Carlinville, *"There isn't a home for rent in Carlinville,"* remarked a real estate agent, after trying to secure a home for a new comer. When one stops to think, that is quite correct. There isn't any new industry, causing condition locally. A number of officers from the CCC camp have moved with their families to this city, also several of the instructors at Blackburn College. Rents for some of the homes, which are not modern, are around $7.50 a month, while the average rent for bungalow cottages is around $20.00 to $25.00 a month. These same houses a year ago were renting for $30.00 to $40.00 a month.

Note: Don't you find it an amazing coincidence that the above article appeared in the paper and then one week later, Standard announced that they were going to sell the homes in Carlinville and Schoper? Standard Oil had been reading the paper all along and read every one of Mr. McClure's articles – they just didn't care. They saw their opportunity to now to sell off the houses and then they would be finished with Carlinville.

I personally love Mr. McClure's writing style. He has a "no holds barred," "get it all out in the open," kind of style. I so need him to be the editor of a newspaper in Carlinville again. He would help me in my fight to get Standard Addition cleaned up. The Carlinville Democrat newspaper always backed me in my endeavors for the betterment of Standard Addition. Of course, this could have been because I worked there for so many years. There have been some good editiors there and my bosses were two of them. Thank you Tom Hatalla and Ed Albracht for all the help, I learned a lot from you guys.

☙ Standard Oil to Dispose of Houses ❧

☙ On **August 29, 1935,** the Carlinville Democrat, said that Standard Oil was disposing of their houses in Carlinville. The article stated, *"It was decided by Standard Oil officials, to dispose of*

Note: Don't you think that they could have used a better word than "dispose?"

their houses, in Standard Addition. The company built 156 houses in ***1918-1919.*** *The houses are all 2-story with basement, containing 5 or 6-rooms. At one time, the houses were almost all occupied by Standard employees who worked in the mines at Berry and Schoper. When the mines closed, some 12-years ago, the families moved from this city and the houses have been empty. There are 153 houses, to be offered for sale. C.W. Clark, local manager, will have charge of the disposition of the houses and the price will range from $350 to $500, which will include the house and the lot. Some repairs, of course, will be necessary."*

☙ Standard Addition Houses Selling Rapidly ❧

☙ By **September 6, 1935,** the houses were selling fast. The Carlinville Democrat reported,

"Houses in Standard Addition, that have been vacant for more than a decade, have been selling at a fast rate and nearly all the homes are sold. People are busy making repairs and cleaning up the area. The activity is a welcome sight and suddenly Standard Addition has come back to life.

Most homes are being bought by persons who plan to make their homes in them, while others are being bought for investment purposes. Among purchasers are families from out of town, who plan to move and make their home here. A drive through the Addition will reveal that new owners are actively on the job bringing their new property into habitable condition. Gone is the deserted air, which has marked this section, and in its place there is activity. Windows are being cleaned, floors scrubbed, steps and roofs repaired and arrangements for papering and painting are in progress. Everywhere workmen are busy. The speed at which these homes are being snapped up is really remarkable and clearly indicates that people anxious to own their own home. The ***Democrat*** *is happy to congratulate the new homeowners. We hope that everyone who buys a home will be glad that he did so, and we believe that Standard Oil's decision to sell these houses will benefit Carlinville."*

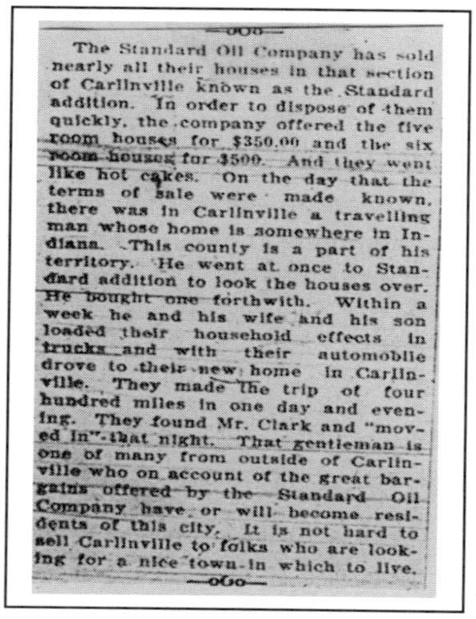

The Standard Oil Company has sold nearly all their houses in that section of Carlinville known as the Standard addition. In order to dispose of them quickly, the company offered the five room houses for $350.00 and the six room houses for $500. And they went like hot cakes. On the day that the terms of sale were made known, there was in Carlinville a travelling man whose home is somewhere in Indiana. This county is a part of his territory. He went at once to Standard addition to look the houses over. He bought one forthwith. Within a week he and his wife and his son loaded their household effects in trucks and with their automobile drove to their new home in Carlinville. They made the trip of four hundred miles in one day and evening. They found Mr. Clark and "moved in" that night. That gentleman is one of many from outside of Carlinville who on account of the great bargains offered by the Standard Oil Company have, or will become residents of this city. It is not hard to sell Carlinville to folks who are looking for a nice town in which to live.

☙ Standard Addition Houses Almost All Sold ❧

☙ On **September 12, 1935**, the paper stated, *"Standard has sold nearly all of their houses in Standard Addition. In order to dispose of them quickly, they offered 5-room houses for $350.00 and 6-room houses for $500.00. On the day terms of the sale were made known, there was in Carlinville, a man from Indiana. He went at once to Standard Addition to look the houses over. He bought one and within a week, he, his wife and son, loaded up their belongings in trucks, and moved to their new home in Carlinville. They made the trip of 400-miles in a day and evening. They found Mr. Clark and moved in that night. That gentleman is one of many from outside of Carlinville, who on account of the great bargains offered by Standard, will become residents of this city. It is not hard to sell Carlinville to folks who are looking for a nice town in which to live."*

☙ Mr. and Mrs. Evan Loveless, who resided on South Broad Street, purchased a Standard Addition home on **September 17, 1935**.

☙ Frieda Pfaff bought a Standard Addition home **September 17, 1935** on Johnson Street.

☙ Mr. and Mrs. Clarence Mayer of Virden, purchased a Standard Addition home **October 3, 1935** and moved into it as soon as repairs were finished.

☙ On **October 17, 1935,** the **Carlinville Democrat** ran a list of deeds that had been recorded at the Macoupin County Courthouse for homes in Standard Addition.

More Standard Addition deeds that were filed at the courthouse were listed in the *Carlinville Democrat* on **October 24, 1935**.

On **January 16, 1936**, Fred H. Smith opened a grocery store in Standard Addition. Virgil Wheeler was the manager.

CONTINUATION OF THE FOREGOING
ABSTRACT OF TITLE
-TO-

LOT NUMBERED SEVEN (7) IN BLOCK NUMBERED TEN (10) IN STANDARD ADDITION TO THE CITY OF CARLINVILLE, IN THE COUNTY OF MACOUPIN AND STATE OF ILLINOIS.

46. EXCEPTING COAL AND MINERALS UNDERLYING SAID PREMISES AND THE RIGHT TO MINE AND REMOVE THE SAME.

FROM THE 28th DAY OF APRIL, A.D. 1920 AT 5 O'CLOCK P.M.

-BY-
THOMAS J. COSTELLO & CO.
Abstracters of Titles.

Standard Oil Company, By:-R. H. McElroy, Vice President, Attest:-F. T. Graham, Secretary, (Corporate Seal) To:-Warranty Deed Albert W. Marko.	Dated September 11th 1935 and Recorded September 23rd 1935 in Book "300" at page 489 Consideration $350.00

47. Conveys:- Lot Seven (7) in Block Ten (10) in Standard Addition to the City of Carlinville, as shown by plat thereof recorded in the Recorder's office of Macoupin County in Large Plat Book "A" at page 59, together with improvements thereon.

Excepting the coal and mineral underlying said premises and the right to mine and remove the same, which are hereby reserved in the Grantor; also subject to an easement in favor of the public over a strip of land six feet in width off the rear end of said premises for the laying and maintaining of sewer, gas and water mains and electric wire lines and poles, as provided by said plat and the dedication thereof.

This deed is made pursuant to a resolution of the Board of Directors of said company, authorizing the same.

Right is the deed of our home when Standard Oil sold it in 1935 for $350.00.

Deeds For Houses In Standard Addition
October 24, 2935

Name	Lot	Block	Price
Henry Schneipp	6	8	$500
Charles E. Dey	4	9	350
Tillie Mitchell	11	7	500
L.C. Wagonblast	14	10	500

Note: The curve where Charles Street enters the Addition, was never taken out, to this day it still curves to the left as you drive past Tom Thumb Park into the Addition.

Deeds Recorded For Houses In Standard Addition
October 17, 1935

Name	Lot	Block	Price
Adolph L. Keune	3	6	$500
Ralph A. Hayes	7	12	500
A.S. Newport	9	12	500
Robert J. Brown	17	7	500
Fred S. Morse	4	8	500
Joe E. Davis	18	12	500
Harry A. Rosenthal	1	7	500
John C. Murphy	0	7	500
Jacob Levora	5	7	500
Kenneth L. Johnson	17	11	500
Ruby Cochran	6	1	350
J.E. Thomae	0	12	500
Joseph Latvaltis	9	8	500
Walter H. Kasten	5	9	350

To Eliminate Curves On Charles

Carlinville City Council met **February 6, 1936** to discuss straightening curves that were on Charles Street at that time. The curves were at the intersection of Mayo and 2nd South Street and at the section where Charles Street enters Standard Addition. The proposal was only discussed. Alderman Monetti proposed, to make Rice Street in Standard Addition a boulevard with stop signs to decrease accidents.

Petition For Gas Service in Addition

On **March 5, 1936**, Carlinville City Council was presented a petition signed by 50 residents in Standard Addition requesting Illinois Power and Light Commission to extend gas service into that section of the city. About 5,900-feet of line were necessary to put the service into the Addition and under the law, the company was required to construct a maximum of 100-feet of line to serve anyone who wished to become a permanent patron. The matter was referred to Carlinville Utility Committee for consideration. The ordinance committee was instructed to draw up an ordinance designating Rice Street a boulevard between North Broad Street and University Street. The Mayor and City Clerk were instructed to prepare a resolution for the improvement of the street under the Motor Fuel Tax Law. A motion prevailed that Alton Railroad be asked to remove the Q-C tracks from High Street.

Note: These tracks were laid to bring the railroad cars carrying the houses into the Addition.

1935-1936 Coldest Winter In Fifty Years

Thomas C. Loehr, local weather observer, reported **March 12, 1936,** the **1935-1936** winter season was coldest in 50-years for the area. From Christmas Day to **February 29,** the thermometer went to zero twice and below 18 times. The ground froze to a depth of 20 to 30-inches. Macoupin Creek froze to a depth of 15-inches while on lakes and ponds it froze from 22 to 28-inches.

Thermometer readings for 20 coldest days in the period.

December 25-------------Zero
December 26---------2 below
December 27---------2 below
January 3------------12 below
January 22-----------12 below
January 23------------5 below
January 24------------5 below
January 25------------3 below
January 26-----------10 below
January 27-----------15 below
January 28-----------12 below
January 30-----------13 below
January 31-----------12 below
February 5------------8 below
February 9------------5 below
February 10-----------6 below
February 11---------------Zero
February 15-----------2 below
February 18---------10 below
February 19-----------6 below

A contract to oil Carlinville streets was awarded to Truman Fiall of Springfield, IL on **June 25, 1936,** to spread 22,000 gallons of oil on streets in the city including: East Street from 1st North to Rice, Buchanan Street from Oak to University, Johnson Street from First North to Burton, High Street from Nicholas to Burton and University from Nicholas to Burton.

From One Extreme To The Other

On **July 16, 1936**, a heat wave in Carlinville beat the 50-year record, according to Thomas C. Loehr, local weather observer. It began on **July 4**.

Thermometer readings since **July 4, 1936.**

July 4-------------103
July 5-------------103
July 6-------------103
July 7-------------105
July 8-------------103
July 9-------------106
July 10-----------106
July 11-----------106
July 12-----------108
July 13-----------109
July 14-----------112

Thermometer readings since **July 23, 1936.**

July 23-----------106
July 24-----------102
July 25-----------105
July 26-----------107
July 27-----------109
July 28-----------100
July 29------------92

Entries in the County Recorder's office at the courthouse on **July 26, 1936:** a deed to Howard Crawford by Standard Oil, lot 8 block 9 in Standard Addition for $350.

An Inspection of Standard Oil mines was made **July 30, 1936,** by Chicago engineers. There were rumors in Carlinville that Standard was going to sell the mines; the fact that the inspection was an appraisal, greatly added to the rumors.

Carlinville Water Supply Almost Gone

The Carlinville Democrat on **September 24, 1936,** said that the Carlinville water supply was almost gone, *"The water situation had reached a critical stage. Considerable quantities had been released from the big lake at Schoper, but unfortunately, this action failed to provide the* *hoped for relief. The old creek itself, was badly in need of a drink and the quantity of water released from the lake was insufficient to create a stream large enough to reach the pumping station."* Citizens resorting to use their wells were cautioned to boil all water before using. Several more releases of water from Schoper Lake held the town over until the water was replenished with heavy rains during the first part of **October 1936.**

☙ The home of Mr. and Mrs. Russell Arnett in Standard Addition was damaged by fire on **October 1, 1936**. The Arnett's were away and their home was discovered engulfed in fire around midnight. Considerable damage was done before the fire was extinguished.

Note: I believe this home was in the middle of the 1000 block of Sumner Street and was torn down as a result of the fire. There has been an empty lot there since the 1930's.

☙ The Carlinville Democrat stated on **October 15, 1936**, *"Standard City will soon have electric lights. Posts were being placed in the main part of the village and other side streets to carry the electricity from the Nilwood lines. Electricity will also be available in the homes."*

☙ Entries in the Macoupin County Recorder's office for deeds on **November 12, 1936,** included: Standard Oil to C.H. Mayer, lot 9 block 1, in Standard Addition for $500.00.

☙ On **November 19, 1936,** Mrs. Robert Davenport entertained friends with a party held at her new home in Standard Addition.

☙ Entries in the Macoupin County Recorder's office for deeds on **December 17, 1936,** included: Standard Oil to Adam J. Schaefer, lot 11 block 9, in Standard Addition for $350.00.

☙ Illinois Power and Light Corporation was nearing completion on construction of running their power lines from Nilwood, IL, to Standard City, IL, on **February 4, 1937.**

☙ Standard City Now Has Electricity ❧

☙ On **February 11, 1937,** the Carlinville Democrat reported that electricity had been connected to Standard City, IL. *"Monday, residents of Standard City were, for the 1st time, able to enjoy electricity. The event was marked by a celebration held in one of the storerooms at Standard City. More than 4-miles of line were constructed, and farmers living along the line were also invited to the celebration. 9 farmers living along the line will avail themselves of the service. In Standard City, 40 patrons were on the line when the juice was turned on."*

☙ Entries in Macoupin County Recorder's office for deeds on **February 11, 1937,** included: Standard Oil to Lewis Calvert, lot 6 block 12, in Standard Addition for $350.00.

☙ At the **April 5, 1937**, Carlinville City Council Meeting, Aldermen Monetti and Fraser suggested resurfacing asphalt streets in the city, most of which were in Standard Addition. Plans were to use Motor Fuel Tax money. The matter was referred to the Street Committee.

Blackburn College Celebrates 100 Years

June 5 through June 9, 1937, Blackburn College celebrated their Centennial. Gideon Blackburn founded the College in **1837.** The celebration lasted through the weekend. A parade with 20-plus floats was the main feature of the celebration.

*Left is the Illinois Power Company Float during the Blackburn College Centennial Parade in **1937** – note Standard Addition houses in the background, this photo was taken at the corner of University and Wilson Streets.*

*Right is the Women's Club float during the **1937** Blackburn College Centennial Parade – note Standard Addition houses on University Street in the background.*

Right is one of many student floats during the Centennial Parade. Again, note Standard Addition houses in the background.

A Special Thank You to the Blackburn College library staff for use of the Centennial Parade Photos and the above photo from their archives.

Blackburn College students at work planting in the field across from Standard Addition in ***1927*** *– again, notice the houses on University Street in the background.*

⸙ Standard Addition Residents Complain ⸙

⸙ A number of Standard Addition residents filed complaints with the Board of Review on **June 24, 1937.** Complaints filed concerned tax assessments being higher for their properties than when they were owned by Standard Oil. The Board of Review's opinion was, that the houses were sold at considerable less than their real value and since improvements were made, the assessed value was raised.

⸙ Approve Improving Addition Streets ⸙

⸙ Through efforts of Alderman Eugene Monetti, a resolution was passed on **July 1, 1937**, for resurfacing of Standard Addition streets. Cost for resurfacing was between $8,000 and $12,000.

⸙ Last of the Houses Sold ⸙

⸙ On **August 31, 1937**, the last unsold house in Standard Addition was sold. It was lot 9 block 9 to L.L. Spears by Standard Oil for $500.00.

⸙ Mrs. Charles Enner entertained at a dinner at her home in Standard Addition on **September 30, 1937**. Guests attending were from Gillespie and Carlinville.

⸙ One of the last houses sold from the Schoper Mine site, was sold to Otto and Lydia Armour in **1938.** Of course, the houses that were sold, had to be removed from the site and what follows is the Armour's house journey to its new home site.

⸙ A House Is Moved ⸙

After the mines closed, houses at Schoper remained, as those in the Addition had, empty. Some were abandoned when people left their homes in search of employment elsewhere.

When Standard sold the houses in Standard Addition, some that were at Schoper were sold as well. Others were just left there for a few years and then sold later. The ones that were sold were moved from the site. Out of the 12 Sears, Roebuck & Co. homes that were built at Schoper, all but 2 were sold and moved. Of the last 2, one eventually fell into such a state of disrepair that it was condemned and destroyed. The other remained the mine superintendent's home until the Schoper Mine was sold and he moved out. The house was then destroyed by fire.

Otto and Lydia Perottit Armour bought the last house that was moved from the site, in **1938.**

The Armour's began moving their home from Schoper in the summer of **1938.**

We believe that the Armour's house is the one the arrow is pointing to in the photo. The house was painted white at a later date. This house is the Langston Model – the same as the Langston Models that are found in Carlinville's Standard Addition.

Right, the Armour's house still sitting on its foundation, just before being moved from Schoper – notice, it even has a Sears, Roebuck & Co. garage in the back yard.

Below, lifting the house onto the wagon.

Right is the D-2 Caterpillar used to pull the house, this photo was taken at Schoper – note another Sears house in the background.

The house was jacked up, loosened off of its foundation and loaded onto a wagon for the move, which consisted of several miles and crossed Macoupin Creek in two places. The Armour's needed something that could pull a lot of weight, so they secured a D-2 Caterpillar to move the house.

Photos left and below are of the house lifted up off of its foundation and ready to roll. You can still see another house beside it at Schoper.

Once the Caterpillar hooked onto the wagon, the house was mobile and ready to go. They chose to use the Standard City – Atwater Road as the moving route. The first obstacle was the Macoupin Creek, east of the A.D. (Dude) Anderson farm, on the Standard City – Atwater Road. When they got to the creek, they had to shore up the area where the creek crossed the road with railroad ties, so that the wagon could go over it without sinking. It was told that when this task was finished, the day was nearly over, so the house sat on the Standard City – Atwater Road overnight. Can you imagine, how long this took to do, not too mention how heavy railroad ties are?

Right are the men shoring up, the creek so that the house could be pulled over it on the wagon.

*The house as it sat along the standard City – Atwater Road in **1938.***

The house on its way across Macoupin Creek.

*Left is drainage tile being laid shortly after the house arrived at the farm site in **1938.***

A Special Thank You to Wendell and Geneva Armour for the photos and history of their special Sears, Roebuck & Co. Honor Bilt Home.

When they resumed house moving, and got the house to Rte 108, they reached another obstacle – again it was Macoupin Creek. The house was too wide to pass between the bridge side railings on 108, so it sat overnight again, until they figured out what to do. They decided to jack up the house even higher so that it would pass above the railings. This idea must have worked, because the house got to the farm where it still sits today. This farm is located about 4-miles east of Carlinville on Rte 108. It is known as Armour's Dairy Farm.

*Left is a new chimney being added to the house right after it arrived at the farm site in **1938.** A basement was dug directly beneath the house later. Stone blocks used for the foundation appear to be the original blocks that the house was built on – these same blocks are used as foundations for most of the homes in Standard Addition.*

*The rear of the house in **1939,** just before Armour's moved in.*

Otto and Lydia passed the farm on to their son and his wife, Wendell and Geneva Armour, in **1963.** Wendell and Geneva remodeled it, and added on an addition, almost the size of the house itself, in **1966.** It was home to them and their children until **1997,** when it passed on to their son and his wife, Todd and Martha Armour, and their four children, Mitchell, Lydia, Ava and Grayson. Todd and Martha still carry on the dairy farming business. Armour's Dairy Farm is the only dairy farm in Macoupin County today.

*Photo below and left is of Wendell and Geneva's children around the kitchen table in the house.(R to L) Greg – 3 yrs old, Dane – 8 yrs old, Gail – 9 yrs old, Wendy – 6 yrs old and Gina – 5 yrs old. This photo was taken in June of **1963** for the Farm Bureau Paper, "A Farmers Guide," to promote milk.*

Milk - A Health Drink For All

Right is a photo of Armour's Dairy Farm as it looks today.

Chapter XI

"It's Mine No More"

Mine No More

HISTORY THAT TOOK ALMOST A DECADE TO MAKE AND ONLY SECONDS TO BRING DOWN.

Tuesday, **August 27, 2002** will be remembered by many people in the area, especially those in nearby Standard City, as the day the landmark they had seen their whole lives driving to and from home came toppling down.

*The Schoper Mine just before its demolition in **2002**.*

The chimney of the 84-year-old Schoper Mine was reduced to rubble by explosives placed in three designated areas within the massive structure. Demolition of the old mine had been in progress for the few weeks prior and the chimney was the only part of the building that required the use of explosives to bring it down. Shortly after 10:00 a.m. the warning sounded and the chimney was gone.

Photo to the left is of the Schoper Mine after the Smokestack was dynamited and the bulldozers started pushing it in.

Below is some of the heavy machinery that was used for the demolition at the old mine site.

The old mine site was reclaimed by the Abandoned Mine Land Department of Natural Resources. A few weeks later, the whole mine building was buried in a large hole that the demolition crew had dug beside it for that purpose.

When the mine was at its full operation it was the largest coal producing mine in Illinois. For many years Standard Oil was a pillar of the Carlinville community and Standard City.

As you can see from the photo to the right, the Schoper Mine site had become an eyesore and was home to a lot of graffiti as well as people and teenagers who shouldn't have been hanging around it.

Schoper Mine closed in **1925.**

Many miners found work in other areas and moved out of Standard City, Carlinville and the area. Some people at the Schoper site near Standard City even moved their homes with them. The mine had been sold a few times over the years.

Schoper Lake had supplied Carlinville with water in a couple of desperate times and remains the only legacy left of Standard Oil at that site. With the old buildings gone, the demolition of the mine complete, and grass planted over the area, it looks as if nothing has ever been there.

Above are my two sons, Jacob (left), Chris,(right), on the top of the debris pile at the old mine site.

My family and I Just couldn't resist going out there to explore after the demolition started. I had to get a couple of souvenirs and they now occupy a proud place in the garden of our back yard in Standard Addition.

Schoper Lake is still full of water and a lot of people frequent it for fishing, there are still boat docks and every now and then you can hear the faint hum of a boat, but mostly it is just a peaceful serene place. Carlinville now has a very stable water supply and will never again need to borrow from the lake. It is all a memory.

Left and below are more demolition photos from the mine site.

It is hard to believe that such a major event ever took place there.

Chapter XII

"Today"

Today

☙ Standard Addition Today ❧

University Street Today.

Even the old Mining Goods Store is still there (although it is not in use now, it has been updated with some work and a fresh coat of paint.)

☙ Standard Addition is today home to a wide variety of occupants, not unlike anywhere else in America. There are those who couldn't care less and then there are those who are proud of the heritage of their home and area. Those of us who are proud, are the ones who suffer through the constant influx of the "I don't care people" who own the houses just to rent them out and never do anything to fix them up, or the renters who rent them, and don't give a darn about the house or the yard. I personally feel that these people should be banned from owning or renting a home in the "largest single collection of Sears, Roebuck & Co. homes in the world." There are those of us who have restored our home to its former grandeur with the greatest of detail and those who have totally updated and rearranged their homes. Either one of these is a beautiful thing to see in the end result; the latter just makes it a little harder to tell what model the home once was.

After tons of complaining and begging, we finally got two small signs to mark the entrances to Standard Addition. I would like to see signs on every corner in the Addition, but that is just a dream at this stage.

Right is one of the 8-inch by 7-inch signs.

Below is Johnson Street in during one of the floods.

There are many beautiful homes here, but for every beautiful one, there is another that, let's just say, is less than beautiful. The old storm sewers that were put in all of those years ago cannot handle large rains and the Addition ends up with flooded streets and basements. When your office is in your basement, as mine is, it can be devastating.

Still, this area remains the perfect place to

More photos of the flooding that sometimes occurs in the Addition. These were taken in front of my house on Johnson Street. My son Christopher was 14-years old in the top photo; he is in the middle of the street.

The bottom photo is the street in front of our home.

raise children; there are always plenty of kids to play with. The park is only a couple of blocks away from every home. The streets are in some places so thick with trees that you can't see the sky through the canopy of leaves. Everybody knows everybody else; it's not long after someone new moves into the Addition before everybody knows them too. Many things could be done to improve Standard Addition, but those things are a dream, a hope that maybe someday…

Right are the 1000 blocks of Johnson Street (top) and High Street today – Lots of trees. Below is the Charles Street curve it was never taken out.

Maybe someday the people who live in Carlinville who call it "just Standard Addition" or "just those Sears Homes," will give it the respect that tourists, the media, magazines, newspapers, T.V. Shows, and the people who live halfway around the world have given it.

Until that day comes, I will not stop fighting for Standard Addition.

Whitehall Models Today

Roseberry Models Today

Carlin Models Today

Gladstone Models Today

Lebanon Models Today

Langston Models Today

Madelia Models Today

Warrenton Models Today

ೞ Unknown Models Today ೨

ೞ Schoper Today ೨

ೞ If you were to visit the Schoper Mine site today, you would see no evidence that a little mining community was there almost a century ago. All of the homes as well as the mine itself have vanished. The dam and the big Schoper Lake, or as others called it, the Schoper Reservoir, are still there. If you happen to catch site of someone fishing in the big lake, then you've seen the only activity in that area these days.

Sometimes, you might see a boat or two on the lake and if you look close, there are a few boat docks dotting the sides.

The area is so serene and peaceful that a person might even catch a nap there.

The mine was so completely razed that it looks as if it had never been there.

There are no bricks, no railroad tracks, not even a nail or a piece of glass.

The site where the Schoper Mine once stood. Today the mine is just a memory.

To the left is Schoper Lake as it looks today.

Below is the dam as it looks today, it is still 85-years later, holding back the water.

Outside view of the dam wall today.

If you do happen to go out for a drive to see the old mine site and Schoper Lake, don't forget your fishing pole.

You can easily find the area by the street sign, which is clearly marked "Mine Rd."

While you're fishing, if you lean back, close your eyes and listen really hard through the sounds of the water hitting the dam wall, you just might hear the faint whistle of the old mine signaling the start of a new workday, somewhere in time.

☙ Standard City Today ❧

Standard City is home to about 100 people today and it is pretty much the same as any other "Small Town U.S.A."

The only businesses that are there are a couple of bars and possibly some home businesses. Most of the people work out of the area, although a few do work in Carlinville. It is a very quiet place and as most small towns go, everybody knows everybody.

Standard City looking east.

Standard City, Standard Addition and some of the people that moved here to work in the mines have simply stayed – they are all a unintentional legacy left to us by Standard Oil Company.

Standard Oil Company would like to consider this an episode in its history, "better off forgotten."

Forget about Standard Addition, Standard City, Schoper, Carlinville, the mines and most importantly, the people who lived in that era?

Never!!

The memories will live on forever.

“Finally”

ଔ Finally ଓ

ଔ We are at the end of this book, but only in the middle of the ***"Standard Addition Saga"*** as I have only taken this history to **1938** and there are many years to go until the present. **1938** to the present will be my next book, which I have already begun. I hope that I have at least put an end to some of the rumors and misinformation about Standard Addition, Schoper, Carlinville and Standard City. If one person says they learned something from this, then it will have all been worth it.

Don't Buy A House There!

ଔ When we moved into the Addition, we were told by our families and friends not to buy a house there because it was considered the ***"bad"*** area of Carlinville. I had always wondered why the older Carlinville citizens detested the Addition so much. While doing research for this book, I found out why.

Standard Oil Company came to this area with their money, with promises of job permanency, with demands and most of all, with promises of a better life.

"Build, build houses and take Care of Our People."

Carlinville did everything that was asked of them, built cinder roads, built houses, built more houses, built even more houses. Carlinville extended the city limits so that the Standard Oil Company Addition would be in the city limits. Carlinville extended the sewer lines, the water lines and paved the streets. Carlinville citizens welcomed them with open arms and practically an open pocketbook. What did Carlinville get in return? A couple of years of a boom that was it's biggest in history and then the boom itself went boom. First, there was too much coal and that only allowed for on again and off again work in the mines. Then, finally, the biggest ***"deboom"*** of all – Standard Oil Company found out that they could buy coal cheaper than they could pay for it to be mined, and the mines closed for good.

MINES CLOSED PERMANENTLY

It came as a shock to Carlinville citizens, but they sucked in their guts and lifted their chins and said, ***"We can survive without the great Standard Oil Company."*** It was hard at first, businesses were closing right and left and people were moving out of the town. Tons of houses were empty; but, Carlinville survived.

Standard Oil Company Owes Carlinville

You bet Standard Oil Company owes Carlinville a huge moral obligation. They lied and did not live up to their end of the bargain. In 9-years time to go from boom to bust and take a whole county, town and its citizens, who believed your word, along for the ride, that is just plain wrong. Then to just pack up and leave one day without even letting Carlinville know what was going on, that was an even greater wrong. To have Carlinville extend sewers, city limits and pave streets in an Addition named after the great Standard Oil Company and then just leave it all to fall down, that was the greatest of wrongs.

The great Standard Oil Company didn't care; they had gotten everything they wanted from Carlinville and then some.

Carlinville still owed and was still paying on all of those improvements that the great Standard Oil Company asked for when that very same great Standard Oil Company took off and left

Carlinville in the red. I don't blame the older Carlinville citizens for being mad; it was their tax dollars that paid for these improvements. Their property tax assessments were raised accordingly and they paid for everything out of their own pockets. Business owners lost their businesses because they were told to, or just plain had to, carry extra stock and merchandise to meet the demands of a booming town. Then, almost as fast as it came, the boom went and many businesses were left with all of that merchandise they couldn't pay for or sell. The older citizens of Carlinville probably heard this very same stuff from their parents, who lived through it.

Yes, it's a fact, the great Standard Oil Company did and does owe Carlinville, but they're never going to own up to that.

There's Only One Place Like it in the World!

But, Carlinville has something, something that the great Standard Oil Company could not take with them - a 9-block area in the northeast part of town containing 152 homes. Not just your average homes, but Sears, Roebuck & Co. Honor Bilt Modern Catalog Homes. There is no other area like it in the world and there never will be. It is a one of a kind rarity, and it is in the city of Carlinville; the Standard Oil Company saw to that. Everybody wants it, but Carlinville is the only one that has it. People, media, magazines, T.V. shows and more all want to see it. People from Japan, Europe and all over come to Carlinville to tour the homes in the Standard Oil Company Addition.

Standard Oil Company officials have said that their episode and activities in Carlinville were ***"Better off forgotten."***

Wake up Carlinville!!! I say forget the great Standard Oil Company!!! Let's turn this Addition into a million dollar a year moneymaker. Make it into a special historic district and adopt special codes and requirements for the people who own houses in it and maybe change the name – it doesn't have to remain Standard Addition. Fix the areas up in it that are bad. Highly publicize it. Open it for tours all of the time. Most of all, make a museum so that all who come may know the history of Carlinville's boom time.

And maybe, just maybe the great Standard Oil Company will regret one thing that they did in Carlinville. Maybe they will regret leaving all of those houses behind and selling them off so cheap. Then Carlinville can sit back and laugh at the great Standard Oil Company.

Ha! Ha! Ha!!!

I work independently from the Carlinville Community Chamber of Commerce.

I still welcome tours; anybody who wants information, help with an article, report or to know if they live in a Sears Home or even just to chat about the houses-

You can reach me by email at ***flori@ctnet.net.***

Bibliography and References

Chapter I.
Standard Oil Company – Oil Pioneer of the Midwest. New York: Published by Appleton Century Crofts, Inc., 1955.

The Carlinville Democrat Newspaper. Carlinville, IL: Published by ALA Enterprises.

The Carlinville Public Library, a public institution. Carlinville, IL

Houses By Mail- A Guide To Houses from Sears Roebuck & Co., a book. Washington, D.C.: Published by The Preservation Press, 1986.

Stanolind Record -1919 -1926 – An employee newsletter. Published by Standard Oil Company.

Wood River History Museum – a public institution. Wood River, IL.

Chapter II.
The Carlinville Democrat Newspaper. Carlinville, IL: Published by ALA Enterprises.

Mary Schroder, an individual, Carlinville, IL.

The Story of Macoupin County 1879-1979, a book. Carlinville, IL: Published by Carlinville & Macoupin County Sesquicentennial, Inc., 1979.

The Macoupin County Historical Society, an organization. Carlinville, IL.

The Carlinville Public Library, an organization. Carlinville, IL

Standard Oil Company – Oil Pioneer of the Midwest, a book. New York: Published by Appleton Century Crofts, Inc., 1955.

Stanolind Record -1919 -1926 – an employee newsletter. Published by Standard Oil Company.

Wood River History Museum – a public institution. Wood River, IL.

Plat Books and County Maps – Macoupin County Courthouse – a government institution. Carlinville, IL

Chapter III.
The Carlinville Public Library, a public institution. Carlinville, IL

The National Library of Congress, an organization. Washington, D.C.

The Carlinville Democrat Newspaper. Carlinville, IL: Published by ALA Enterprises.

The Springfield News Record Newspaper. Springfield, IL: Published by the Springfield News Record.

Quality Title & Abstract Company, an organization. Carlinville, IL.

Standard Oil Company – Oil Pioneer of the Midwest. New York: Published by Appleton Century Crofts, Inc., 1955.

Houses By Mail- A Guide To Houses from Sears Roebuck & Co., a book. Washington, D.C.: Published by The Preservation Press, 1986.

Plat Books and County Maps – Macoupin County Courthouse – a government institution. Carlinville, IL

Stanolind Record -1919 -1926 – An employee newsletter. Published by Standard Oil Company.

Wood River History Museum – a public institution. Wood River, IL.

Chapter IV.

The Carlinville Democrat Newspaper. Carlinville, IL: Published by ALA Enterprises.

The Carlinville Public Library, a public institution. Carlinville, IL

The Story of Macoupin County 1879-1979. Carlinville, IL: Published by Carlinville & Macoupin County Sesquicentennial, Inc., 1979

Sears, Roebuck & Co. Honor Bilt Modern Homes Catalogs, Chicago, IL: Published by Sears Roebuck & Co. 1908-1940.

Stanolind Record -1919 -1926 – An employee newsletter. Published by Standard Oil Company.

Wood River History Museum – a public institution. Wood River, IL.

Plat Books and County Maps – Macoupin County Courthouse – a government institution. Carlinville, IL

Chapter V.

The Carlinville Democrat Newspaper. Carlinville, IL: Published by ALA Enterprises.

The Carlinville Public Library, a public institution. Carlinville, IL

Standard Oil Company – Oil Pioneer of the Midwest. New York: Published by Appleton Century Crofts, Inc., 1955.

Sears, Roebuck & Co. Honor Bilt Modern Homes Catalogs, Chicago, IL: Published by Sears Roebuck & Co. 1908-1942.

Houses By Mail- A Guide To Houses from Sears Roebuck & Co., a book. Washington, D.C.: Published by The Preservation Press, 1986.

Stanolind Record -1919 -1926 – An employee newsletter. Published by Standard Oil Company.

Wood River History Museum – a public institution. Wood River, IL.

Plat Books and County Maps – Macoupin County Courthouse – a government institution. Carlinville, IL

Chapter VI.

The Carlinville Democrat Newspaper. Carlinville, IL: Published by ALA Enterprises.

The Carlinville Public Library, a public institution. Carlinville, IL

Blackburn College Archives, Blackburn College, a learning institution. Carlinville, IL.

Quality Title & Abstract Company, an organization. Carlinville, IL.

The National Library of Congress, an organization. Washington, D.C.

Sears, Roebuck & Co. Honor Bilt Modern Homes Catalogs, Chicago, IL: Published by Sears Roebuck & Co. 1908-1940.

Debra Spangler-Whitlock, an individual. Carlinville, IL
Stanolind Record -1919 -1926 – An employee newsletter. Published by Standard Oil Company.

Wood River History Museum – a public institution. Wood River, IL.

Thomas Hatalla, an individual. Carlinville, IL.

The Chicago Tribune Newspaper. Chicago, IL: Published by the Chicago Tribune, Inc.

Sears, Roebuck & Co. Plumbing Catalogs, Chicago, IL: Published by Sears Roebuck & Co., 1920's.

Sears, Roebuck & Co. Modern Heating Catalogs, Chicago, IL: Published by Sears Roebuck & Co., 1920's.

Plat Books and County Maps – Macoupin County Courthouse – a government institution. Carlinville, IL

Chapter VI-1.
Sears, Roebuck & Co. How To Build Your Readi Cut Home, an instruction manual. Chicago, IL: Published by Sears Roebuck & Co., 1927.

Chapter VII.
The Carlinville Democrat Newspaper. Carlinville, IL: Published by ALA Enterprises.

The Carlinville Public Library, a public institution. Carlinville, IL

Standard Oil Company – Oil Pioneer of the Midwest. New York: Published by Appleton Century Crofts, Inc., 1955.

The Internet, Internet Explorer, general search for topic.

The Bloomington Pantagraph Newspaper. Bloomington, IL: Publisher Unknown.

The Jerseyville Democrat Newspaper. Jerseyville, IL: Published by the Alton Evening Telegraph Newspaper.

Mary Schroder, an individual, Carlinville, IL.

The Macoupin County Courthouse Centennial Book, a book. Carlinville, IL. 1969.

Blackburn College Archives, Blackburn College, a learning institution. Carlinville, IL.

Stanolind Record -1919 -1926 – An employee newsletter. Published by Standard Oil Company.

Wood River History Museum – a public institution. Wood River, IL.

Plat Books and County Maps – Macoupin County Courthouse – a government institution. Carlinville, IL

Chapter VIII.
The Carlinville Democrat Newspaper. Carlinville, IL: Published by ALA Enterprises.

The Carlinville Public Library, a public institution. Carlinville, IL

Standard Oil Company – Oil Pioneer of the Midwest. New York: Published by Appleton Century Crofts, Inc., 1955.

Stanolind Record -1919 -1926 – An employee newsletter. Published by Standard Oil Company.

Wood River History Museum – a public institution. Wood River, IL.

Plat Books and County Maps – Macoupin County Courthouse – a government institution. Carlinville, IL

Chapter IX.
The Carlinville Democrat Newspaper. Carlinville, IL: Published by ALA Enterprises.

The Carlinville Public Library, a public institution. Carlinville, IL

Standard Oil Company – Oil Pioneer of the Midwest. New York: Published by Appleton Century Crofts, Inc., 1955.

The Chicago & Alton Railroad Company, a transportation institution.

Stanolind Record -1919 -1926 – An employee newsletter. Published by Standard Oil Company.

Wood River History Museum – a public institution. Wood River, IL.

Deeds and Title Transfers - Macoupin County Courthouse – a government institution. Carlinville, IL

Chapter X.
The Carlinville Democrat Newspaper. Carlinville, IL: Published by ALA Enterprises.

The Carlinville Public Library, a public institution. Carlinville, IL

Standard Oil Company – Oil Pioneer of the Midwest. New York: Published by Appleton Century Crofts, Inc., 1955.

Quality Title & Abstract Company, an organization. Carlinville, IL.

Wendell and Geneva Armour, husband and wife as individuals. Carlinville, IL.

Blackburn College Archives, Blackburn College, a learning institution. Carlinville, IL.

Deeds and Title Transfers - Macoupin County Courthouse – a government institution. Carlinville, IL

Chapter XI.
The Carlinville Democrat Newspaper. Carlinville, IL: Published by ALA Enterprises.

Standard Oil Company – Oil Pioneer of the Midwest. New York: Published by Appleton Century Crofts, Inc., 1955.

Deeds and Title Transfers - Macoupin County Courthouse – a government institution. Carlinville, IL

Chapter XII.
The Carlinville Democrat Newspaper. Carlinville, IL: Published by ALA Enterprises.

Sears, Roebuck & Co. Honor Bilt Modern Homes Catalogs, Chicago, IL: Published by Sears Roebuck & Co. 1908-1940.

The Story of Macoupin County 1879-1979. Carlinville, IL: Published by Carlinville & Macoupin County Sesquicentennial, Inc., 1979

Sears, Roebuck & Co. Catalogs, etc. used for images and information within this book. Items are from the personal collection of Laurie A. Flori.

1906 – The Steel Square and Its Uses - Book.
1908 – Sears, Roebuck & Co. Honor Bilt Modern Homes Catalog.
1910 – Sears, Roebuck & Co. Home Builders Catalog – Reproduction Catalog.
1914 - Sears, Roebuck & Co. Honor Bilt Modern Homes Catalog.
1917 – 18x24 Fold Out Brochure for Sears, Roebuck & Co. Honor Bilt Farm House, Barn and other buildings.
1917 – Sears, Roebuck & Co. Catalog Ad for the Carlin Model House.
1917 – Instructions In Modern Carpentry - Book.
1917 – Sears, Roebuck & Co. Honor Bilt Modern Homes Catalog.
1918 – Sears, Roebuck & Co. Honor Bilt Modern Homes Catalog.
1918 – Sears, Roebuck & Co. - Advertising Postcard for their Honor Bilt Modern Homes.
1918 – Sears, Roebuck & Co. - Advertising Postcard for the Lebanon House Model.
1918 – The Steel Square and Its Uses – Book.
1918 – Instructions In Modern Carpentry – Book.
1919 – The Steel Square and Its Uses Volume 1 and Volume 2 – Books.
1919 – Sears, Roebuck & Co. Catalog Ad for the Warrenton and Madelia Model House kits.
1919 – Sears, Roebuck & Co. Honor Bilt Modern Homes Catalog.
1919 – Sears, Roebuck & Co. Catalog Ads for modern plumbing.
1919 – Sears, Roebuck & Co. Catalog Ads for modern lighting fixtures.
1919 – Sears, Roebuck & Co. Catalog Ads for hardware, millwork and other fixtures.
1920 – 18" x 24" Fold Out Brochure of Sears, Roebuck & Co. Honor Bilt Modern Homes.
1920 - Sears, Roebuck & Co. Standard Built Modern Homes Catalog.
1920 - Sears, Roebuck & Co. – Interiors In Home Furnishings Brochure.
1921 - Sears, Roebuck & Co. –Electric Light and Power For Country Homes Catalog.
1922 – Sears, Roebuck & Co. Honor Bilt Modern Homes Catalog.
1925 - Sears, Roebuck & Co. Honor Bilt Modern Homes Catalog.
1926 - Sears, Roebuck & Co. Catalog of Houses - Reproduction Catalog.
1926 - Sears, Roebuck & Co. Honor Bilt Modern Homes Catalog.
1927 - Sears, Roebuck & Co. - Building Material, Millwork and Roofing Catalog.
1927 - Sears, Roebuck & Co. Modern Plumbing and Heating Systems Catalog.
1928 - Sears, Roebuck & Co.–Leather Bd – 75-Pg Instruction Man. - How to Build Your Readi-Cut Home.
1929 - Sears, Roebuck & Co. – How to Install Modern Plumbing Systems Catalog.
1930 - Sears, Roebuck & Co. Catalog - Houses Designs of the Thirties – Reproduction Catalog.
1930 - Sears, Roebuck & Co. Honor Bilt Modern Homes Catalog.
1931 - Sears, Roebuck & Co. – Honor Bilt Building Materials Catalog.
1935 - Sears, Roebuck & Co. Honor Bilt Modern Homes Catalog.
1937 - Sears, Roebuck & Co. Honor Bilt Modern Homes Catalog.
1939 - Sears, Roebuck & Co. Honor Bilt Modern Homes Catalog
1940 - Sears, Roebuck & Co. - Homart Lighting Catalog.
1950 - Sears, Roebuck & Co. Book – Catalogs and Counters.
1986 - Sears, Roebuck & Co. Book –100th Anniversary 1886-1986
1998 – Homes In A Box – Catalog Reprint – Contains Sears, Roebuck & Co. Honor Bilt Homes.